European Security in the 1990s

Deterrence and Defense after the INF Treaty

CNSS
Center for National Security Studies
Los Alamos National Laboratory

ISSUES IN INTERNATIONAL SECURITY
Series Editors: Robert E. Pendley and Joseph F. Pilat

BEYOND 1995: The Future of the NPT Regime
Edited by Joseph F. Pilat and Robert E. Pendley

EUROPEAN SECURITY IN THE 1990s: Deterrence and Defense after the INF Treaty
Walter Laqueur and Leon Sloss

A Continuation Order Plan is available for this series. A continuation order will bring delivery of each new volume immediately upon publication. Volumes are billed only upon actual shipment. For further information please contact the publisher.

European Security in the 1990s

Deterrence and Defense after the INF Treaty

Walter Laqueur

Center for Strategic and International Studies
Washington, D.C.

and

Leon Sloss

Leon Sloss Associates
Washington, D.C.

With the assistance of Philipp Borinski

Foreword by
David M. Abshire

PLENUM PRESS • NEW YORK AND LONDON

Library of Congress Cataloging-in-Publication Data

European security in the 1990s : deterrence and defense after the INF Treaty / Walter Laqueur and Leon Sloss.
p. cm. -- (Issues in international security)
Includes bibliographical references.
ISBN 0-306-43442-3
1. Europe--Defenses. 2. North Atlantic Treaty Organization--Armed Forces. 3. Nuclear weapons--Europe. I. Series.
UA646.E9255 1990
355'.03304--dc20 89-49472
CIP

This series of books has been prepared as an account of work sponsored by the Los Alamos National Laboratory. Neither Los Alamos National Laboratory, any agency thereof, nor any of their employees makes any warranty, expressed or implied, or assumes any legal liability or responsibility for the accuracy, completeness, or usefulness of any information, apparatus, product, or process disclosed, or represents that its use would not infringe privately owned rights. Reference herein to any specific commercial product, process, or service by trade name, mark, manufacturer, or otherwise does not necessarily constitute or imply its endorsement, recommendation, or favoring by the United States Government or any agency thereof. The views and opinions of authors expressed herein do not necessarily state or reflect those of the United States Government or any agency thereof.

A Division of Plenum Publishing Corporation
233 Spring Street, New York, N.Y. 10013

Printed in the United States of America

The Center for National Security Studies

The Center for National Security Studies (CNSS) is a studies and analysis organization of the Los Alamos National Laboratory. Drawing on the broad knowledge at Los Alamos of science and engineering relevant to national security issues, the Center's research focuses on the interaction between technology and policy and on developing insights that may improve the relationship between the development of new technology and the achievement of national policy goals.

The principal mission of the Center is to promote and conduct long-term research and analysis in the broad areas of defense policy and arms control, focusing in particular on the requirements for technology that derive from trends in national and international security policy. In addition, it helps Los Alamos management and technical staff better address national defense needs by providing insight into national security policy, strategy, and technology issues, and by assisting in Laboratory long-range planning. The Center also provides a forum for the exchange of ideas on international security issues among Laboratory personnel, government agency staffs, university faculties, and interested citizens.

To implement its programs and activities, the Center supports an interdisciplinary staff, drawing on expertise from the social sciences, including history and political science; the physical and biological sciences; engineering; and mathematics. The Center conducts an active visitor and consultant program. It encourages short- and long-term visits by experts from government and private institutions and by university faculty members.

The Center sponsors seminars, workshops, and conferences designed to bring Los Alamos personnel into contact with outside experts and to stimulate broader discussion of the relationships between technology and U.S. defense policy. The Center also acts to enhance communication between Los Alamos and other organizations, such as colleges and universities, that are studying issues of interest to the Laboratory.

The publications program of the Center is intended to communicate the results of studies to Laboratory personnel and to reach a wider policy analysis community in government, military, and academic circles, as well as private industry. Central to the publications program is this book series, *Issues in International Security*. The volumes in the series are based on research conducted by the Center's staff and by internationally recognized experts working with the CNSS. A primary goal of the series is to promote the reasoned analysis of international security issues, with emphasis on how these issues shape and are shaped by technological developments.

Foreword

Rarely if ever have the political, economic, and military foundations of the Western Alliance been in such a state of flux. Walter Laqueur and Leon Sloss, therefore, deserve credit not just for the quality of the analysis in this superb book but also for the timeliness of its appearance.

As Laqueur says in his chapter "Touring the Western European Defense Horizon," if the likely development of European defense policies is not particularly reassuring, at least it gives no grounds for despair. The list of problems we face is a daunting one. First there is the spiralling cost of defense expenditures, particularly in the absence of significant NATO or inter-European cooperation. This is particularly serious in light of the reluctance to increase, or even maintain, current expenditures in the midst of Mikhail Gorbachev's "peace offensive" and the extraordinary changes in Eastern Europe, both of which have had a dramatic impact on Western public opinion.

There is also a problem in the perceived relative economic decline of the United States vis-à-vis Western Europe and Japan, which only exacerbates calls to reduce the number of American troops in Europe. Other dangers to the political cohesion and military credibility of the alliance include demographic trends that threaten current manpower levels, transatlantic acrimony over the burden-sharing issue, and political pressures (particularly in West Germany) toward denuclearization and even neutralism.

To address all of these elements, and more, requires a mixture of vision, strategy, and political skills. As Laqueur says, "a new strategic concept is needed both for the individual countries and for NATO as a whole."

Even though the situation with West Germany is radically new, it has a precedent in the mid-1960s. Back then, General de Gaulle lost faith in NATO's nuclear strategy and in U.S. dominance, declared that the USSR was no longer a threat, and withdrew from the military side of the alliance. It all sounds suspiciously like Germany today. But in response to de Gaulle, NATO acted admirably by producing the Harmel Report, which articulated the policy of defense and dialogue that has been NATO's guiding star ever since.

Then, while I was at NATO in 1983, we had a different kind of crisis. Following the stalemate caused by the Soviet walkout from negotiations, an East–West study set up by Leo Tindemans laid the groundwork for NATO's successful pursuit of the "Treaty between the United States of America and the Union of Soviet Socialist Republics on the Elimination of Their Intermediate-Range and Shorter-Range Missiles."

The momentum of the Cold War was broken when the Soviets tried and failed to split NATO on the issue of deploying intermediate-range nuclear weapons. In response, in a masterful diplomatic move, Moscow reversed course. If Stalin had created NATO by his threats, Gorbachev would unmake it by removing that threat. As a result of the General Secretary's deft diplomatic maneuvering, NATO moved from victorious unity in 1983–1984 to potentially dangerous disunity.

NATO's current disarray means that despite the promise of change in the Soviet Union, there is new peril in Europe. As the USSR is confronted by destabilizing forces—some springing from its own "new thinking"—the West confronts its own forms of destabilization. The result is a transformation of Europe in which the Soviet threat has diminished but the dangers of instability have increased. In the 1990s, with the Soviet empire crumbling, Eastern Europe moving to liberate itself, intensified interest in German reunification, the withdrawal of American troops, transatlantic protectionism, and economic conflict, the Western security system could disintegrate. At worst, this could lead to the types of uncertainties and miscalculations that produced World War I. Even in a less pessimistic scenario, however, there will still be a breakup of the Cold War order, and a questioning of some of NATO's fundamental tenets that accompanied that order. It is now necessary for NATO to follow these earlier precedents but with a more comprehensive mandate to develop a NATO grand strategy for the 1990s.

The first aspect of this should be an attempt to do what NATO has resisted doing before, namely, examining the critical triangular relationship between contentious short-range nuclear forces (SNF), improvements in conventional forces, and negotiations for nuclear and conventional arms control. If short-range nuclear forces have any use, it is to offset, however crudely, the Warsaw Pact's conventional superiority. But if the conventional balance can be stabilized through a combination of NATO force improvements and conventional arms control agreements, then the requirements for modernizing SNF could be reconsidered.

We have, therefore, the makings of an alliance deal. The American proposal, which makes SNF negotiations contingent on conventional reductions and insists in advance that SNF will not be reduced to zero, identifies some of these trade-offs correctly. However, what is still lacking is any link between SNF and the third side of the triangle, namely, improvements in NATO's conventional forces. Consequently, there is a pressing need for an overall net assessment of our critical vulnerabilities, caused not just by superior Soviet forces but also by flagrant weaknesses deriving from NATO's own maldeployments, ammunition shortages, and so forth.

One of this book's most valuable insights is its observation that NATO must develop systems "that can adapt to weapons perhaps not yet fully conceived." And although the authors do not identify such systems specifically, an excellent example of how we might close our self-created vulnerabilities would be the development of the most important Pentagon program of the past decade—"competitive strategies." Instead of matching the Warsaw Pact tank for tank, the concept emphasizes the incorporation of smart and brilliant weapons into new weapons systems and tactical concepts that might turn the tide of a conventional battle in NATO's favor. As well as rendering NATO's defense posture more effective, competitive strategies also provide an uncontroversial alternative to the deeply divisive short-range nuclear missiles. Political unity within the alliance is more important than any weapon system.

Discussions within the alliance at every level—from summit meetings to "public diplomacy"—are needed on a huge range of issues in order to convince Western electorates of the need to retain a credible defense posture. Walter Laqueur and Leon Sloss are fully aware that security involves not just military matters but politics and economics as well. It is this fundamental awareness providing a backdrop to their remarkable expertise and grasp of the details that makes this book such a welcome contribution to the debate.

David M. Abshire
President, Center for Strategic and International Studies
Former Ambassador to NATO

Preface

A specter is once again haunting Europe—the specter of profound political and economic change. These changes offer promise as well as peril, and their ultimate effects are as yet unknown. Their implications for European security have been debated by politicians, policy analysts, pundits, and publics throughout Europe and the United States, and have created or exacerbated differences within the Atlantic Alliance. Contentious issues such as burden-sharing and short-range nuclear force modernization, along with the challenges of Mikhail Gorbachev's public diplomacy, pose a threat to the future of the alliance. And NATO's responsiveness to the political explosion in Eastern Europe, with its implications for East-West relations, will pose a fundamental challenge to the continued importance of the alliance for European security.

Will the Atlantic Alliance survive the challenge posed by the dramatic changes now occurring in Eastern and Western Europe? Are there viable alternatives to NATO? How will Europeans define and ensure their security in the years ahead? Will the apparent lessening of the Soviet threat preclude the need for costly and politically divisive decisions on defense budgets and force improvements?

The distinguished authors address these and other questions about European defense. They give particular attention to economic and demographic trends and to the attitudes of publics and political elites in the major countries of Western Europe. They also examine the implications of the Intermediate Nuclear Forces Treaty for the future of NATO; identify trends toward denuclearization and their meaning for European defense, which ultimately has relied on U.S. nuclear weapons for more than four decades; and explore the viability of alternative defense concepts for future European security. The compelling analysis provided in the volume is supported by an appendix containing an extensive collection of documents pertaining to European security.

In compiling the documents presented in the appendix, we acknowledge and appreciate the assistance of the United States Arms Control and Disarmament Agency; the United Nations Department for Disarmament Affairs; the British Information Services, New York; and the German Information Center, New York. We also acknowledge and appreciate the dedication and contributions of Center for National Security Studies staff. Arthur Nichols, the Center's editor, established with the publisher editorial procedures and schedules, and he copyedited the volume. Molly Cernicek's contributions were substantial—she played a critical role in collecting, preparing, and organizing the documents in the appendix. The skilled assistance of Janis Dye was indispensable in preparing the manuscripts for the publisher. And Wanda Franks deserves special recognition for facilitating our communications with the authors.

JOSEPH F. PILAT
ROBERT E. PENDLEY

Contents

Appendix: European Security in the Postwar Period: Documents from 1948–1989

Introduction

From the time of its founding forty years ago, the North Atlantic Treaty Organization (NATO) has depended on U.S. nuclear weapons for deterrence and, ultimately, defense. Once it was recognized that the force objectives required for the conventional defense of Europe put forward at NATO's 1952 Lisbon Conference could not be realized for political and fiscal reasons, NATO adopted a strategy that relied in the last instance on the threat of "massive retaliation" by U.S. strategic forces.

The credibility of this doctrine was seriously undermined by the Soviet deployment in the 1950s of intercontinental bombers and intermediate-range ballistic missiles (IRBMs) targeted on Western Europe, and by their development of intercontinental ballistic missiles (ICBMs). And, since the late 1950s, the United States has had to reassure the Europeans of the credibility of extended deterrence by developing and demonstrating a range of strategic nuclear options and by deploying nuclear weaponry in continental Europe or in adjacent waters so as to establish the "coupling" of the defense of the European theater to the U.S. strategic deterrent. Accordingly, during the 1950s and 1960s, NATO undertook the forward basing of U.S. nuclear-capable *Thor* and *Jupiter* medium-range ballistic missiles (MRBMs), which were removed in 1963, and F-111 medium-range fighter-bombers in Europe, and the United States assigned *Polaris* submarine-launched ballistic missiles (SLBMs) to the Supreme Allied Commander, Europe (SACEUR). As well, in response to the same political and military pressures, a host of proposals were put forward at this time for a NATO nuclear force. The only such proposal that was not stillborn was the Multilateral Force (MLF), but it too was ultimately unrealized.

Nevertheless, European concerns over the credibility of the American nuclear guarantee and transatlantic differences over the role of nuclear weapons in the alliance persisted. In 1967, in response to problems of NATO's nuclear strategy, the alliance adopted a doctrine of "flexible response." A fragile compromise between European interest in absolute deterrence and assured coupling to U.S. strategic nuclear forces, and American interests in avoiding an immediate recourse to a strategic exchange with the Soviets in the event of conventional conflict in Europe, the doctrine of flexible response has been official NATO doctrine since it was adopted.

If questions about the credibility of NATO doctrine and the need for intermediate-range nuclear forces (INF) had been muted in the 1950s and 1960s because of U.S. nuclear superiority at both the strategic and theater levels, developments in the 1970s radically altered the Eurostrategic landscape. During the 1970s, not only did the Soviet Union achieve parity with the United States in central strategic systems, but with modernization of Soviet intermediate-range forces, and in particular the introduction of SS-20 IRBMs and *Backfire* bombers, it attained nuclear superiority in the European theater. The immediate response of the United States was to deploy an additional wing of F-111s to the United Kingdom and to increase the number of SLBM warheads assigned to SACEUR. However, under pressure from the Europeans, and especially the Federal Republic of Germany (FRG), the necessity for NATO to develop a comparable INF capability, in order to preserve the viability of a flexible response strategy that depends upon a capacity for "escalation control," was acknowledged.

NATO'S DUAL-TRACK DECISION

That NATO's theater nuclear forces should be modernized was recognized as early as 1975, in Secretary of Defense James Schlesinger's report on *The Theater Nuclear Force Posture in Europe*. Schlesinger's successor, Donald Rumsfeld, pursued a series of modernization proposals that followed this report in meetings of NATO's Nuclear Planning Group (NPG) in 1975 and 1976. In May 1977, as part of its Long-Term Defense Program (LTDP), NATO established Task Force 10, under the NPG, to undertake a review of remedial measures required in the alliance's posture on INF. At the NPG meeting held at Bari, Italy, in October 1977, a High Level Group (HLG) directed to address the need for theater nuclear force modernization, and the technical, military, and political implications of various modernization proposals, was established at the urging of the FRG, whose growing concern about the Eurostrategic balance in the age of superpower strategic parity and the strategic arms limitation talks was eloquently stated in Chancellor Helmut Schmidt's speech to the International Institute for Strategic Studies in October 1977.

After two years of deliberations in the HLG, during which the discussion of options came exclusively to focus on long-range theater nuclear forces, that is, INF, a fragile consensus was created within the alliance. On this basis, NATO's Council of Ministers decided on December 12, 1979, to modernize NATO's nuclear deterent through the deployment of 572 INF systems, including both *Pershing* IIs and ground-launched cruise missiles (GLCMs). These systems were to be deployed with American forces, under U.S. command and control. The United States was to finance the development and production of these nuclear systems, while the European allies were to contribute to the costs of their infrastructure and support through NATO's Infrastructure Fund. It was also recommended in the communiqué of

December 12, 1979, that U.S.–USSR arms control negotiations should include certain INF systems and that the United States should propose to the Soviet Union that such negotiations should begin in the near term. Finally, the unilateral withdrawal in 1980 of one thousand nuclear warheads based in Europe was announced.

Military Rationale

NATO's INF modernization decision was motivated by military and political factors, and its rationale reflects both of these considerations. Militarily, the modernization was necessitated by developments on the strategic and theater nuclear levels: first, by the advent of strategic parity between the United States and the Soviet Union, which threatened the credibility of the U.S. guarantee to utilize strategic nuclear forces in order to provide extended deterrence to the European members of NATO; second, by the shift from NATO to Warsaw Pact theater nuclear superiority, as the NATO long-range nuclear systems (U.S. F-lll s and British *Vulcans*) were becoming obsolete and less capable of penetrating improved Soviet air defenses, and the Soviet nuclear capabilities in the European theater increased dramatically by a modernization program that included the introduction of long-range systems such as the SS-20 and the *Backfire* bomber. In addition to Soviet actions that effectively eliminated NATO nuclear superiority at the strategic and theater levels, the Warsaw Pact also increased its conventional forces and enhanced the superiority it has enjoyed in that sphere since the 1940s.

As a consequence of these developments, the credibility of NATO's nuclear forces, which are designed along with its conventional forces to preserve peace by deterring or defeating aggression, was threatened. NATO's nuclear strategy of flexible response was perceived to be increasingly inadequate to fulfill its traditional military roles, that is, to deter a conventional attack on Europe by Warsaw Pact forces or the Pact's first use of nuclear weapons, or to defeat a conventional Warsaw Pact invasion if NATO's conventional defenses falter. Concern about the viability of NATO nuclear doctrine became apparent as the Soviet Union appeared to be developing the means for "escalation dominance" at a level for which NATO had no credible response (i.e., INF), in light of doubts about the capacity and willingness of the United States to utilize its strategic forces in response. Ultimately, the INF modernization was held to enhance NATO's deterrent by avoiding the establishment of escalation dominance by the Soviet Union at the INF level, and thereby buttressing the strategy of flexible response. As for the latter, the INF modernization could be expected to reduce European concerns about the link between NATO's theater nuclear forces and U.S. strategic nuclear forces, and about the viability of extended deterrence, that is, coupling, which also appeared threatened by developments in the 1970s. Depending upon perspective, the deployment could be seen either to increase or to decrease the prospects for a nuclear war limited to Europe. Certain military aspects of the decision, including the rejection of the sea-based option, were designed precisely to allay European concerns about the U.S. commitment to the alliance.

Political Rationale

If the military rationale of the INF modernization could be construed as compelling, political concerns and considerations were also crucial in the evolution of the December 1979 decision. First, as indicated, Europeans were concerned about the direction of the SALT negotiations, particularly their treatment of "gray-area weapons" such as cruise missiles, and their implications for the United States' commitment to the alliance. And the decision may have reflected concerns about the continued political existence of the alliance as well as about a Soviet military threat. This is evidenced by the decision that the deployment be land-based and visible in order to ensure coupling, and extensive enough to ensure the viability of flexible response but not to establish a Eurostrategic balance that would be suggestive of decoupling. Other political considerations came into play that reflected concerns about European domestic opinion, including the necessity of tying the INF modernization proposal to an offer to initiate arms control negotiations, and the requirement that the FRG not be the only nonnuclear-weapon state within NATO to accept deployment.

NATO'S Nuclear Dilemma

The dual motivation of NATO's dual-track INF deployment decision created ambiguities, primarily because the public debate surrounding the proposed deployment focused for political reasons on the Soviet SS-20 threat. If the deployment was presented as a counter to the deployment of SS-20s because this was simpler than recourse to a NATO nuclear doctrine that represented a rather fragile compromise of European and American interests that were once again at the point of serious divergence, this created a belief that arms control in and of itself could resolve problems associated with NATO's nuclear strategy and further Western security interests. After all, if the deployment of INF were only necessary to counter the SS-20, the reduction or elimination of the SS-20s would remove the rationale for introducing *Pershing* IIs and GLCMs. If, however, the modernization of INF was necessitated by the obsolescence of certain NATO nuclear systems, or the necessity of new, more capable systems to revitalize NATO nuclear strategy, then the arms control track of the 1979 decision had a much more limited role. Ambiguity about the motivations of the 1979 decision consequently resulted in difficulties in assessing the priority of the two tracks, with the potential for losing sight of the original and enduring military reason for the modernization. Indeed, the political logic of the INF modernization decision in light of European perceptions of international and domestic developments resulted in a nuclear dilemma for NATO.

TOWARD THE INF TREATY

At the opening of INF negotiations in November 1981, after the Reagan administration's "zero-zero" proposal, which was rejected by the Soviets in February 1982, the prospects for an agreement did not appear promising. When it was learned that the United States and the Soviet Union both rejected the July 1982 "walk in the woods" compromise developed by Paul Nitze and Yuli Kvitsinsky, which would have allowed both sides 75 launchers and 150 nuclear-capable aircraft, all hope appeared to have been lost, and the deployment itself for a time appeared in jeopardy. It was widely believed that the Soviets would not negotiate seriously until the political will and the solidarity of the alliance was demonstrated by the implementation of the deployment decision. On the other hand, it was believed that the political conditions in certain NATO countries would make deployment problematic without some significant demonstration of movement in the Geneva negotiations.

To a large extent as a consequence of political problems surrounding NATO's decision to deploy modern INF in the European theater, there was at this time a renewed, at times riotous, public debate over NATO nuclear strategy and the necessity of nuclear weapons for NATO's defense. In this context, there were calls for the adoption of a doctrine of "no first use" and of an effective conventional defense that would make use of new technologies and new operational doctrines to take advantage of Warsaw Pact vulnerabilities and NATO strengths in order to provide an effective and affordable defense of Europe. There were also mass demonstrations throughout Europe—the "peace protests" that preceded the initial INF deployment and peaked during the "hot autumn" of 1983, which was not nearly so warm as had been predicted. Yet, INF deployment proceeded apace. For all of the major European members of NATO, implementation of the deployment track of the dual-track decision appeared to have demonstrated the unity and political cohesiveness of the alliance, and to have restored some of the credibility of U.S. leadership in NATO and of the U.S. nuclear guarantee in which it is grounded.

This reasoning was reaffirmed by progress on the arms control track following the deployment. The Soviets walked away from the INF talks in late 1983, as the deployment began. But after a year away from the negotiating table, and apparently recognizing that they had blundered in terms of their public diplomacy, they decided to return. After discussions between U.S. Secretary of State George Shultz and Soviet Foreign Minister Eduard Shevardnadze in January 1985, U.S.–Soviet negotiations on INF resumed in March 1985, in the context of broader Nuclear and Space Talks (NST) that also included strategic nuclear forces and defense and space issues. During 1985 and 1986, from the opening of the NST to the Reykjavik summit, it had appeared that the sides might agree on INF reductions to equal, lower limits. In the final communiqué of the Geneva summit, President Ronald Reagan and General Secretary Mikhail Gorbachev supported the idea of such an interim INF accord.

However, at the beginning of 1986, Gorbachev, in a major move toward the U.S. position, proposed the elimination of all INF in Europe. On January 15, 1986, Gorbachev called for nuclear disarmament by the year 2000, proposing for intermediate-range nuclear forces the adoption and implementation of a decision on the complete elimination of medium-range missiles of the Soviet Union and the United States in the European zone—both ballistic and cruise missiles—as a first step toward ridding the European continent of nuclear weapons. Although the language is less than revealing, it would appear that this Soviet INF offer was tied to a freeze on British and French nuclear forces and a provision that would preclude Britain's purchase of the *Trident* II SLBM. After putting forward the INF offer, Gorbachev stated that at the same time the United States should undertake not to transfer its strategic and medium-range missiles to other countries, while Britain and France should pledge not to build up their respective nuclear arsenals. In a meeting with Senator Edward Kennedy in Moscow only weeks later, Gorbachev indicated that an INF agreement was possible without agreement on defense and space arms, asserting that the only conditions the Soviets placed on their proposal to eliminate INF in Europe were the freeze on British and French forces and the agreement not to transfer strategic and theater systems to any other country. In Reykjavik in October 1986, the sides agreed in principle to eliminate all INF based in Europe.

While the question of the linkage of the INF negotiations to strategic and in particular to defense and space issues, which had appeared earlier, reappeared once again after the Reykjavik summit, the Soviets eventually did not persist in their demands. Moreover, the Soviets decided to offer a concession and to accept the global ban on INF preferred by the United States, as well as an agreement to eliminate shorter-range nuclear systems (with ranges between 500 and 1000 kilometers). This "double-zero" solution was embodied in the Treaty between the United States of America and the Union of Soviet Socialist Republics on the Elimination of Their Intermediate-Range and Shorter-Range Missiles, which was signed by President Reagan and General Secretary Gorbachev in Washington on December 8, 1987.

The treaty, like the Reykjavik summit before it, was not greeted enthusiastically in Europe. As had happened so many times in the past, European fears of decoupling and superpower condominium were aroused. The double-zero was seen as possibly initiating a fundamental transformation of the military structure of the alliance with military and political ramifications that were uncomfortable for the Europeans. The INF Treaty could not be rejected for political reasons, but it was not perceived adequately to reflect and protect European security interests.

IN THE AFTERMATH OF THE INF TREATY

For the alliance, the expenditure of political capital associated with the deployment decision has been enormous, given the limited military capability involved. In the aftermath of the INF Treaty and its double-zero, the military policy and posture of the alliance will require reassessment. However, the political controversies that have surrounded the dual-track process have severely constrained NATO's future nuclear options. Modernization of NATO's remaining nuclear forces—as agreed at Montebello in 1983, along with further unilateral reductions of European-based U.S. nuclear warheads—is certain to be politically difficult, as is clearly revealed by the current debate over short-range nuclear force (SNF) modernization. As differences over SNF suggest, NATO's views on nuclear strategy are divided;

NATO's flexible response doctrine has been subjected to such intense scrutiny in the last several years that its credibility and utility may have been compromised.

There is undoubtedly a need to eliminate the imbalance between NATO's and the Warsaw Pact's conventional forces, especially as intermediate-range and shorter-range missiles are eliminated pursuant to the terms of the December 1987 Washington Treaty. This is recognized not only in the United States but also by the Germans, the French, and the British. However, European interests and perspectives pose serious limits to this approach for NATO. The perceived reduction of the Soviet threat, and the dramatic events in Eastern Europe, along with alliance-wide demographic and budgetary problems, and the prospects for a conventional arms accord have greatly complicated the prospects for NATO conventional force improvements. Clearly, arms control is seen in Europe as a far more attractive and, it will be argued, affordable alternative to increases in defense spending and cooperation—even more so after Gorbachev's December 7, 1988, United Nations speech, in which he announced unilateral cuts in Soviet conventional capabilities and his subsequent announcement of unilateral SNF cuts.

A new forum for reducing conventional arms within Europe from the Atlantic to the Urals—the talks on Conventional Forces in Europe (CFE)—has been convened. These negotiations have been accorded a high priority in both the East and the West, and they may lead to militarily significant reductions. The recent proposals put forward by President Bush, which were incorporated into the recent NATO negotiating proposal tabled in Vienna, have greatly reduced the differences between the two sides in the CFE talks. Although the problems remain formidable, the political explosion in Eastern Europe would appear to make a CFE agreement virtually inevitable, albeit less important. To the extent that arms control is seen as the primary solution to European security problems, there may be a drift toward neutralism and Finlandization, as Europeans may seek to ensure and expand the economic and political benefits of détente, and their own security, by this and other nonmilitary means in a rapidly changing world.

Even as arms control is being embraced by Europeans, sporadic interest in European defense options has been revived and might become more serious in the medium and long terms, but the obstacles are formidable. Already the British and French have reportedly been discussing the joint development of air-launched nuclear weapons, and the Germans and French have begun to increase their defense cooperation. France will undoubtedly continue to be in the forefront of such thinking, and the recently "resuscitated" Western European Union (WEU) will probably continue to be a preferred forum of the French.

In the light of these trends, which may undermine NATO's military rationale and capabilities, will the alliance be able to survive? Despite flirtation with the WEU, could this organization, or any newly formed European entity, provide a viable alternative to NATO? How will Europeans define and ensure their security in the 1990s and beyond? Will economic and demographic trends, the challenges of Gorbachev's public diplomacy, or changes in Eastern Europe redefine or undermine European defense in the decades ahead? Will, on the other hand, a perceived lessening of the Soviet threat preclude the need for costly and politically divisive decisions on nuclear and conventional force modernization and, perhaps, even preclude the need for NATO?

In the essays that compose this volume, the authors attempt to address these and other questions about European defense in the years ahead. In "Touring the Western European Defense Horizon," Walter Laqueur examines the political and economic context of the defense of Europe in the foreseeable future, with special emphasis on economic and demographic trends and on the attitudes of publics and political elites in the major countries of Western Europe. Then, Laqueur and Philipp Borinski, in "Western European Defense: The Next Ten Years," look at prospects for post-INF European security, especially the likelihood of European defense cooperation, and the viability of frequently heard proposals for alternative or non-provocative defenses in Europe over a decade span. Leon Sloss, in "Future Requirements for Theater Weapons in Western Europe," addresses the strategic, doctrinal, and other factors that will directly affect future requirements for theater nuclear weapons in Europe, and he examines the implications of the INF Treaty and trends toward denuclearization for the prospects of modernizing short-range nuclear forces and, more broadly, for the future viability of extended deterrence.

JOSEPH F. PILAT

1

Touring the Western European Defense Horizon

INTRODUCTION

A look at present-day European defense policies and their likely development in coming years is not particularly reassuring, but it gives no grounds for despair. This judgment is nowhere clearer than in the Federal Republic of Germany (FRG), where future developments will be of pivotal importance for Europe and the Atlantic Alliance. While sudden, radical changes emanating from within the FRG are unlikely, there is the possibility of slow, creeping changes which would have the same impact on the alliance, albeit in the long run. And the recent changes in East Germany, which are accelerating the debate over German "reunification," have potentially profound implications for the FRG's relations with Western Europe and NATO.

Domestic developments in the FRG and the other countries of Western Europe are critical—the defense consensus, defense expenditure, and the politics of defense strategy will depend on the internal balance of power in the various countries. The principal problem besetting all countries of Western Europe is the spiralling cost of defense expenditure and popular reluctance to increase defense spending. Some of these problems could be solved, or at least some of the potential shocks to defense absorbed, if there were closer NATO or inter-European cooperation, but prospects for successful actions in this direction are not very good. Another problem shared by the European members of the alliance is a demographic pattern which is highly unfavorable for the maintenance, let alone the expansion, of current manpower levels. A new strategic concept is needed both for the individual countries and for NATO as a whole, so that they will be able to cut their coat according to their cloth; only then will it emerge whether the amount of cloth will be sufficient for any coat.

However, a new overall security concept is needed for reasons which have nothing to do with financial situations of, and with demographic trends in, the

European NATO states. As a result of the Intermediate-Range Nuclear Forces (INF) Treaty, various new dilemmas have emerged, and differences of interest have come to light, or have been exacerbated, between the United States and the European allies on one hand, and between European governments on the other. Future agreements with the Soviet Union will make it even more imperative to rethink the nuclear future of NATO. As this report shows, much thought, although often inconclusive or unfeasible, has already been given to this topic in various countries, and the debate is bound to continue.

WEST GERMANY AND THE ATLANTIC ALLIANCE IN THE YEARS TO COME

What will be West Germany's place in the Atlantic Alliance in the near future? For a long time this country was the star pupil in the class of European NATO members. More recently, doubts have been widely voiced about the reliability of West Germany in view of loud and widespread opposition to NATO and to U.S. policy and, generally speaking, the apparent breakup of the national consensus on defense and foreign policy. To address the questions that have been asked in this respect, it is necessary to take the German domestic scene as the starting point.

The Political Parties

The last national elections in West Germany took place in 1987. The Christian Democratic Party (CDU) incumbents led by Chancellor Helmut Kohl lost more than 4 percent of their share of the vote, but since the Social Democratic Party (SPD) also lost votes, the CDU remained the strongest party, and they provided the leader of the government coalition. The next general elections are due not later than January 1991; an earlier date is possible if the Free Democratic Party (FDP), the junior partner in the government coalition, should decide in favor of a new alliance with the Social Democrats, with whom the FDP cooperated for long periods during Germany's postwar history. Political trends since the last general elections have shown a decline in support for the Christian Democrats, a trend which has manifested itself in regional electoral defeats. This has to do with a variety of domestic political considerations, especially in the social field. It could also be that a belief that it is time for a change plays a certain role. The "Greens," who emerged within a short time in the early 1980s as a major force in German politics, seem to have peaked some time ago; the bickering among the leadership of the party and the vagueness of its program (except on the local level) had a detrimental effect on its political standing.

Will NATO be able to count on West Germany in the future, as it has in recent decades? It appears—to put it in a nutshell—that there is no cause for alarm in the short run but that there are certain disquieting factors if one assumes a long-term perspective. Changes have taken place in the foreign political orientation of the Social Democrats in recent years. But even if they should win the next elections (or those following them), they will probably not have an absolute majority; they never had such a majority in any election in the past. Nor is a coalition with the neutralist Greens likely to result in a workable majority. The Social Democrats and the Greens rely more or less on the same electoral reservoir. If, against expectations, the Greens should add to their votes in the next elections, it will be to the detriment of the Social Democrats, and vice versa.

The Social Democratic leaders understand that while in opposition they can advocate all kinds of radical policies, but they also know that they have to go to the next elections with a moderate platform under a leader who is likely to appeal not only to the left-wing militants but who is also popular among the public at large.

If these observations offer a short-term perspective, it is also true that Germany will face political and other difficulties in fulfilling its duties as a key member of NATO in the years to come. Nor, in a longer perspective, can it *a priori* be ruled out that there may be a gradual change in orientation as new political generations come to the fore in the Federal Republic. To assess these developments, a closer look at the parties and, through their prism, at domestic German politics is necessary.

The Christian Democrats

The Christian Democrats are a coalition between various heterogeneous elements and interests: left and right of center, North and South German, Catholic and Protestant, the interests of high finance and industry, as well as a substantial working-class element. The combination of such disparate elements is the source of its strength—its integrative power in being able to offer something to everyone—but also of its weakness. The party has been beset by almost permanent internal strife, and Chancellor Kohl in particular has been accused by his critics of failing to show qualities of leadership. If another obvious leader were in sight, his days would be numbered, but no such generally acceptable figure is available at the present time. The disastrous start in office of Defense Minister Manfred Wörner all but wrecked his career. His successor, Rupert Scholz, was a controversial and ineffective defense minister during his short term in office. Gerhard Stoltenberg, the new defense minister, is widely respected for his competence and integrity but seems to lack massive political appeal. Yet, with all its internal rifts, the CDU, as past experience has shown, always closes ranks at a time of decision; there is too much at stake for everyone. There is much reason to believe that this will also be the case in the future.

The CDU has been on the whole the staunchest proponent of Atlanticism in Germany. There has been occasional criticism of specific American strategies (e.g., Kurt Biedenkopf and certain "Gaullist" trends among the Bavarian Christian Social Union—CSU), but the party has been loyal to the Western Alliance in all decisive situations. If the relationship with America has been a marriage of convenience, rather than a passionate affair of the heart, this is also true with regard to other NATO members. The alliance has been based on the sober realization that Europe cannot today be defended without American help and guarantees, a foundation which could be reevaluated in light of the changes in Eastern Europe, especially in East Germany.

The CDU increasingly stresses specific German concerns and interests having to do with specific economic and demographic constraints. Thus, CDU spokesmen have emphatically rejected "selective deterrence" as well as the doctrine of a "fireproof wall" (Volker Rühe). They have argued that from the German point of view it makes no difference whether the country is destroyed by intermediate- or short-range nuclear missiles, hence the German emphasis on the need for nuclear deterrence rather than defense, and the opposition on the part of all parties against short-range nuclear systems as well as nuclear artillery. About the special relationship with East Germany, the CDU has been less sanguinely optimistic than the SPD, but the changes in the German Democratic Republic may alter the CDU's stance on this issue. In any event, it cannot afford to appear less patriotic than its rivals as far as German unity is concerned. Lastly, there has been marked disappointment with Europe. The high hopes attached to European unity have faded and so has the willingness of the West Germans to act as the paymaster of Europe. The appearance on the political scene of the Republican party, a right-wing group combining xenophobia and anti-Westernism, has further weakened the CDU and introduced an element of uncertainty in German politics which did not exist before.

The Social Democrats

The Social Democrats have undergone important changes since the abdication of Helmut Schmidt, which was largely the result of inner party intrigue and the loss of power in Bonn. The old right wing of the party has been pushed aside and even isolated, and the left wing has been greatly strengthened. Hans-Jochen Vogel, the party chairman, and some other leading figures who originally belonged to the party's center-right faction have moved to the left. The left-wing drift does not manifest itself primarily in social and economic radicalism. There has been no greater labor militancy than before; at present there are fewer strikes than in past years. The radicalization shows itself at present more in foreign policies and attitudes toward defense than in domestic affairs.

The new leadership deplores the "excessive anti-Communism" of their predecessors; it stands for closer and friendlier relations with the Soviet Union, the Soviet bloc, and East Germany. It regards the Soviet Union as a "security partner" of West Germany in maintaining world peace and criticizes U.S. foreign policy almost all along the line, from intervention in the Third World to the Strategic Defense Initiative (SDI). There are frequent complaints about Germany's alleged status as a vassal of America.

If the whole SPD leadership were united in such attitudes, the prospects for U.S.–German cooperation would be bleak. However, even the left wing of the SPD is aware of Germany's dependence on U.S. protection—at least for the time being. (It could well be that this feeling of dependence adds to the resentment.) Not a single Social Democratic leader demands neutrality here and now. They insist on direct negotiations with Soviet block leaders, for instance, on the removal of all chemical weapons from Central Europe. They want to return to détente in Europe irrespective of Soviet activities in other parts of the world. They argue that the Pentagon grossly exaggerates Soviet military power, and they maintain that nuclear weapons could and should be safely removed from Europe.

Many of them have claimed for years that the two global powers are equally to blame for world tensions, and, since Mikhail Gorbachev's approaches to Western Europe stressing "our common home," some have argued that America is more to blame in this respect. But so far no one has insisted on the immediate dismantling of NATO, nor is such a demand likely to appear in the near future. Indeed, the reaction of many SPD politicians, including Egon Bahr, to the revolutionary changes in East Germany has been to insist that the two alliances in Europe be maintained as a measure of stability in the near term. Some American analysts have reached the conclusion that the radicalization of the SPD is a reaction to Reagan foreign policies and is, therefore, likely to change at some future date as U.S. policy becomes more moderate.

But criticism of America was no less pronounced when President Carter was in the White House and Helmut Schmidt acted as West German Chancellor. A strong U.S. foreign policy is as likely to be criticized as a weak one seeking accommodation, just as a weak dollar will be as much a cause of concern to Europeans as a strong one. While Reagan policies have been bitterly criticized by sections of the German population, it is also true that, according to the polls, confidence in America as an ally reached a low in 1979, before Ronald Reagan had been nominated as the Republican candidate.

The radicalization of Social Democracy manifested itself in a declaration by Johannes Rau, one of the more moderate party leaders, in which even he advocated the revocation of the double-track decision, which meant the stationing of missiles in Germany. But he was to retract his statement several weeks later; such declarations are frequently made in the belief that they are necessary to maintain party unity. Responsible party leaders know that, when in power, they could not possibly challenge basic NATO decisions without jeopardizing Germany's position in the Western Alliance. When Oskar Lafontaine was a leader of the party's left wing, he advocated a partial German withdrawal from NATO. As he emerged as a candidate with real chances in the race for the party leadership, his proposals were silently dropped. While many Social Democrats may think that West German and American security interests are gradually diverging, they also know that they will have to rely on the Western Alliance for a long time to come. But if the Social Democrats should win the next elections, and if they do not live up to their promises, there is a danger that they will find the left wing of the party in strong opposition—a constellation similar to the one which brought about the downfall of the last SPD government under Helmut Schmidt.

The Greens

A move by governing Social Democrats back to the center on defense issues could also provide a new lease on life to the Greens, the only party opposed in principle to NATO and to defense in general. The Greens have not been doing remotely as well during the last two years as many observers of the German scene previously expected. After their progress on both a regional basis and countrywide, it was widely expected in 1981–1982 that this movement would soon become the "party of youth" and a vital third force in German politics. Instead, they have been losing votes in regional elections (such as in Hamburg), and their general loss of momentum caused a sharpening of the conflicts between *Fundamentalos* and *Realos* which existed in their ranks even during their meteoric rise. The position of the fundamentalists in the party leadership has been strengthened—at the cost of a loss of influence among the "basis." Their party congresses have shown a near total lack of agreement—indeed an inability to agree on anything at all.

The *Realos* advocated active cooperation with the Social Democrats, whereas the *Fundamentalos* believe that such cooperation would corrupt and ultimately ruin the party. Ironically, at the very time when the fortunes of the Greens began to decline, they received the opportunity to enter into coalitions with the Social Democrats, in view of the inability of the latter to form regional governments without a coalition partner. However, the example of Hesse, where the SPD–Ecologist alliance broke down after protracted disputes has acted as a deterrent to such coalition building.

Such cooperation is bound to bring about a split among the Greens, whose policy is grist to the mills of the CDU, which argues, not without justification, that the Greens' policy is hostile to industry and will cause a decline in standards of living. Thus, Social Democratic cooperation with the Greens on a regional basis is bound to damage them on the national level. The Social Democrats have to present the image of a solid, responsible party, and the rapprochement with the Greens has exactly the opposite effect.

If the fortunes of the Greens are likely to decline, and if the Social Democrats are bound to suffer as the result of cooperating with them, they and their ideas are not likely to disappear soon from the German political landscape. There is a protest potential of between 10 and 20 percent in Germany that will manifest itself in demonstrations, some perhaps violent. And, partly in view of the great publicity given to the phenomenon by the media, and partly in view of the fact that many of the discontented are young (and hence considered to be the future of the nation), these protests will be heard in Germany and beyond her borders. The attention given to them is out of proportion to their numbers and intrinsic importance, but they do exist, their voice is loud, and the phenomenon is bound to have a negative impact both in Germany and among her neighbors, putting in doubt Germany's reliability as an ally.

The Free Democrats

The policies of the FDP can no longer be understood in terms of the traditional conflict between right-wing liberals (North German) and left liberals (South German). Some of the latter, predominant in the 1970s, have dropped out of the party or at least out of its leadership following the party's turnaround under Hans-Dietrich Genscher in 1982. The right-of-center liberals are now in command, and economic issues seem to represent their priorities. In this respect, paradoxically, the FDP today occupies the right of the German political spectrum; it is the only party in the FRG with a neo-liberal, or even neo-conservative, persuasion. It tends to question the welfare state, the sacred cow of all mainstream political forces in West Germany, including the vast majority of the CDU–CSU.

With respect to traditional liberal causes in the social and cultural fields and also in foreign affairs (e.g., détente), the FDP has retained the traditional liberal outlook. The combination of economic conservatism, sociocultural progressiveness, and moderation in foreign policy appeals to a growing number of voters: well-to-do women; German yuppies; and the new middle classes in general. Increasingly, the FDP represents what some have called the Gary Hart factor in West German politics. Its leader during the middle 1980s, Martin Bangemann, embodied the image of a dynamic manager with a modern social outlook. In the "post-Marxist" age, this party might also increasingly appeal to intellectuals. Given all this, if it gets its political act together, the FDP might grow, although not by very much, at the expense of the other major parties.

The foreign policy initiatives of Genscher have been on occasion a matter of controversy between the coalition partners, but little inner-party opposition to his line is recognizable. This line stresses "predictability" and "responsibility," and continued détente "from a firm foothold in the Western camp." It also puts greater emphasis on disarmament and closer relations with the Soviet Union than the CDU approach. But, by and large, foreign policy issues do not represent the main preferences of FDP leaders and voters. However, a residue of the pro-reunification sentiment of the 1960s, often coupled with ill-disguised indifference to the FRG's Western connection, lives on and could be strengthened by the developments in East Germany. Conceivably, therefore, the FDP could accommodate possible neutralist sentiments among mainstream voters in the future. But it is more likely that the ideological pull of the FDP's neo-liberalism will keep the party, despite occasional grumbling, oriented to the West.

Trends Affecting German Defense Policy

Demographic Trends

The West German contribution to NATO will depend to a high degree on available manpower. Defense Minister Stoltenberg has proposed a reduction in total strength of the *Bundeswehr* to 420,000 by the mid-1990s,[1] down from the present strength of 495,000 for all ranks, including 223,000 conscripts (not to count a further 750,000 reserves who could be mobilized in a matter of days). During the late 1970s, the *Bundeswehr* faced certain manpower problems stemming mainly from the increasing unwillingness of young Germans to do military service. The situation has improved; at that time it was sufficient to be transferred to civilian service if a young German informed the draft board by means of a postcard. Since then, the conscientious objector has to appear in front of a committee to justify his decision. Permission is still usually granted, but he has to do civilian service for a period *exceeding* that of normal military service by one-third. This is on the assumption that a recruit to the *Bundeswehr* has to do reserve duty for many years after his release from active duty. Furthermore, growing youth unemployment has probably also helped to swell the number of volunteers.

However, the German government, confronting manpower problems again, had planned to prolong military service to eighteen months, effective from June 1,

1989. The reason is, very briefly, that the West German birthrate has declined since 1964, and it is now (with the exception of the Christmas Islands) the lowest in the world. Since 1973–1974, there has been an *absolute decline* in the number of ethnic Germans. According to EEC projections made in 1980, the number of young German males aged fifteen–nineteen, which was 2.6 million in 1980, will have declined to 1.8 million in 1990 and 1.5 million in 1995. According to a projection undertaken by the ministry of the interior, the number of young Germans (under twenty) will decline from 14.4 million in 1985 to 7.2 million in 2030. In other words, the number of young Germans will decline much faster than the population at large. The population of West Germany rose from 55 million in the 1950s to 62 million in 1973–1974. This was mainly the result of immigration, first from East Germany (2 million between 1955 and 1961), and later on of "guest workers." (Part of this pattern has reappeared with 350,000 ethnic Germans coming from East Germany in 1989.) Today, at least one birth in ten, and probably more, is of foreign parents who are not German citizens and whose offspring will not therefore serve in the German army. The relative part of foreign citizens among the general birthrate will further increase, at least for a decade to come, and this makes the problem which has been indicated even more acute.

Population projections are always unreliable as far as the ups and downs of the birthrate are concerned; the rise of the birthrate in France after the Second World War, after a long period of stagnation, is a well-known example. But even if the German birthrate should again substantially increase at some future date, this will not affect the critical manpower situation of the *Bundeswehr* in the 1990s. The strength of the *Bundeswehr* will substantially decline unless certain measures are taken to alleviate the situation, such as increasing the percentage of professional soldiers, including the need for more young Germans to sign up for longer periods of service, a call-up of married young men, and the call-up of more reservists each year. If these measures are taken, which seems likely, by the late 1980s, this would put another 100,000 young men at the disposal of the *Bundeswehr*. The planned extension of military service from fifteen to eighteen months, which would add another 40,000, has been suspended partly because the higher than expected number of draftees currently available has made this step unnecessary.

Economic Prospects

In 1965 the Federal Republic spent 4.3 percent of its GNP on defense; since 1973 it has been spending consistently 3.3–3.4 percent—less than Britain and France but slightly more than Italy. But by 1987, defense expenditure fell to below 3 percent. In recent years German defense spending has been less than planned by NATO (annual growth of 3 percent) or indeed by the Bonn government. Whereas the growth rates in 1980 and 1981 were 6 percent and 8.2 percent, respectively, they had decreased to less than 2 percent in 1987. The allocations for defense in the state budget decreased to 18.7 percent, and they were likely to decline even further, but for a substantial financial infusion which, given the international constellation and the domestic balance, seemed very unlikely. (It is interesting to note in passing that defense spending of 18.1 percent had been considered essential by the Social Democratic opposition at its last party congress.) The *Bundeswehr* budget was cut in 1988 by DM 10 billion. Since expenses for German army personnel are increasing and will continue to increase both in terms of wages and social benefits, the cuts will affect, above all, acquisition of new weapon systems.[2]

To what extent are economic difficulties likely to affect West Germany's contribution to NATO in the years to come? The West German economy has shown steady growth throughout its existence; only in four years (1967, 1975, 1981, and 1982) was there negative growth. In 1988 the growth rate was 2.5 percent. While the rate of growth in the 1950s was 8 percent, it declined to 5 percent in 1960–1965 and 4.7 percent in 1965–1970. There was a further decline in the 1970s comparable to that in all other industrialized countries, particularly after the oil shock of 1973. According to *Prognos,* a Swiss-based think tank, the West German economy will continue to grow at an average rate of 2.5 percent to the end of the century. If this rate of growth can be maintained, it should be possible to maintain the German contribution to NATO defense without major efforts. However, much depends on the political pressures on the basis of which the decisions on domestic spending are made, and how much priority is given to defense in comparison to other needs at a time when the threat of conflict in Europe is perceived to be rapidly diminishing. It is also true that certain dangers confront the *Bundesrepublik* which could affect its economic performance. West Germany depends more than any other major industrialized country on exports; in 1982 it exported goods and services estimated at 35 percent of the GNP. To give but one example, 60 percent of all German cars and trucks are sold abroad. During the last twelve months the German trade balance (current account balance) has shown a surplus of $64 billion, which was one of the highest ever. But Germany is more exposed than any other country to the ups and downs of world trade. West Germany is vulnerable not only with regard to its ability to compete on world markets, but it is also vulnerable to possible import restrictions in other countries and to the world economic situation in general.

While the inflation rate is at an all-time low, German ability to compete will be restricted by high labor costs; fringe benefits such as various insurances and retirement schemes now amount to 80 percent of wages, and this is unlikely to go down. German industrial development has shown weakness during the last decade as far as the high-technology sectors of the economy are concerned; it also did not try as hard as its competitors to modernize its plants. Of late, there has been a marked improvement in this respect—greater capital investment has gone into high technology such as computers and robots. More money is now spent on research projects than before; the number of new high-venture businesses was one-third higher in 1984 than in the year before. The optimism of 1984–1986 was in stark contrast to the pessimism of 1973–1983, and modernization will undoubtedly bear fruit in the years to come. But with all these encouraging indications, the German economy will remain vulnerable unless internal demand for its products substantially increases, and in view of a shrinking population, the prospects in this respect are not good.

Furthermore, the conservative fiscal policy of the German government—the decision to raise taxes in 1989—has slowed down German performance to some extent. The strength of the deutschemark is also bound to endanger German exports to a certain degree; hence, the International Monetary Fund prediction for economic growth of less than 2 percent in the next few years. It remains to be noted that Germany's own forecasters, who on the whole do not tend toward excessive optimism, for once take a more cheerful view.

It is also true that West German dependence on exports makes it somewhat vulnerable to trade with the Soviet Union and Eastern Europe. Soviet (2–3 percent) and East European imports from West Germany have always been small both in absolute and relative terms; since 1984 they have suffered a substantial decline. (German exports to the Benelux countries are four to five times as much as to the East.) Nevertheless, Soviet purchases are of some importance for certain branches of the German economy (e.g., machine tools). Furthermore, there is the undying hope among some German industrialists that even though the volume of trade is now low, the potential market is big and exports to the Soviet Union will ultimately rise. Constant disappointments over a period of sixty years have done little to dampen this enthusiasm, and the hopes about supplying "oil for the lamps of China" is likely to be a certain factor, albeit not a decisive one, in German Eastern policy. Hence, the decision in 1988 to extend to the Soviet Union a credit line equivalent to more than $2 billion.

A social phenomenon which is likely to cause unrest is the high rate of unemployment—7.8 percent, almost as high as in France and higher than in the United Kingdom. Furthermore, for reasons which are not entirely clear, there has been little decline in German unemployment in recent years, in contrast to other major and most minor countries. A substantial decline is unlikely in the near future because the strengthening of the modern labor-intensive sectors of the economy does not create new jobs. The rate of youth unemployment in comparison with general unemployment is lower in Germany than in other industrialized countries—mainly as the result of the traditional German schemes of apprenticeship. In the final analysis, the availability of places of work and not the training schemes will be the decisive factor, and in this respect a drastic change is likely only if economic output should grow at a much faster rate than projected or if there should be job-sharing on a new, radically different basis. Both eventualities are unlikely in the near future.

In summary, future German governments will find it increasingly difficult to maintain the rate of defense spending.[3] This has partly to do with the relatively sluggish rate of German economic growth, but mainly with the lower domestic priority given to defense in an age of new détente and rising social expenditure.

Anti-NATO Feeling among the Younger Generation

Germany's reliability as a partner in the Western Alliance depends above all on the political mood and the general attitude of the successor generation—the young voters and political activists of the late 1980s and 1990s. A strong case has been made in favor of the proposition that there has been a radical reevaluation of values during the last fifteen years among young Germans—that the rejection of family, work, patriotism, and other values is likely to continue. In foreign politics this changed mood has manifested itself in pacifism, anti-Americanism, the wish to leave NATO, the big demonstrations in 1982–1983 against the stationing of the Euromissiles, occasional terrorist acts, the emergence of the Green party, and the antidefense position taken by leading circles in the Protestant church. According to this school of thought, German youth is anxiety ridden, with a slogan of "no future," or, alternatively, prone to take violent action. It has also been stressed that a substantial percentage of young Germans have expressed negative or skeptical views with regard to military service. The more highly educated they are, the greater their reservations. However, even sophisticated public opinion polls are likely to lead to mistaken conclusions unless interpreted in a wider context. The number of active terrorists in Germany is less than fifty. There is no more anti-Americanism among young Germans than in other European countries. American mass culture is all-pervasive in West Germany as in other European countries; cultural anti-Americanism is by and large restricted to the avant-garde of the extreme left and right. Nor is it a new phenomenon; it was widely observed even in the last century.

The same is true with regard to the longing for neutrality. Many Germans would like to be in a position comparable to Switzerland. This, in any case, was the result of polls during the 1950s; only 37 percent said in 1957 that they actually liked Americans. Sympathies rose during the 1960s, and support for NATO increased to about 70 percent.

This trend was reversed in the 1970s, following the radicalization in schools and universities, and it is certainly true that during this period the attitude of the young generation toward the Western Alliance was more skeptical than that of their predecessors. But, it now appears that it was mistaken to believe (as many observers did at the time) that the change in mood and attitude during the 1970s ("post-materialism," "new social movements") was irreversible. The INF deployment passed without any cataclysmic consequences, despite the dire predictions of some analysts. On the contrary, among the youngest age group a significant change occurred in 1984–1985, inasmuch as only 26 percent maintained that the likelihood of an attack against Germany had increased as the result of stationing American medium-range nuclear missiles. The year before, in comparison, 54 percent had expressed such views. According to recent studies on the political orientation of German youth, such as the German Shell investigation, the number of those afraid of dying in a nuclear war has declined from 50 to 34 percent over the last four years.

Some sanguine observers have drawn the conclusion from these trends that there is no German youth problem—that support for NATO is overwhelming and solid. But this is again an undue simplification of a complex situation. West Germany will increasingly face differences of interest with her allies in the economic as well as the political and military sphere. Much of the pro-European zest has gone out of German politics, and as the movement toward European unification has run out of steam, it has been only natural that the Germans, like their neighbors, have put their own immediate interest above the common ties. Bonn has ignored appeals from Washington and other Western capitals to do more to boost its economy. The differences in strategic approach have been mentioned; increasingly, Germans have argued that they are in a unique position (not only in view of the historical link with East Germany) and that the country is also more vulnerable. Such vulnerability could lead to growing pressure for a German nuclear deterrent, in view of the fact that protection by her allies has become less credible. The other, stronger temptation is to go nonnuclear, which, in practical terms, would mean neutralist. All kinds of alternative "nonprovocative" defense schemes have been mooted; none of them has merit, but the very fact that a great many people have tended to support proposals ranging from the impractical to the absurd points to the existence of a certain malaise. Thus, it has been said that the 1990s could be for Germany a decade of frustration; conscious of its own economic power, it may face foreign political and strategic choices which are none to its liking. Progress toward reunification will more likely be the result of developments in East Germany and elsewhere than decisions in West Germany. In other words, Germany may be no more master of its own destiny than in past decades, and this could lead to political instability both at home and in relations with other countries.

The mood of the younger generation at the present time can be summarized, in brief, as follows. Generations follow each other now more quickly than in the past. The revolutionary avant-garde of 1967 had little in common with the adolescents of the late 1970s, and those at present in their late teens by no means accept the cultural and political guidance of the preceding generation. In view of the rapidly changing moods, it is impossible to predict future orientations. Extreme change in mood and orientation is one of the characteristics of youth. The generation of 1967 entered the "long march through the institutions" in the 1970s and, as the result of their exposure to reality, many of them have substantially retreated from their erstwhile extreme positions. They are now in their late thirties or forties, and from among this age group the political leaders of the 1990s will be chosen. Nevertheless, their outlook differs in some essential respects from previous generations as represented by Konrad Adenauer and Kurt Schumacher, by Kohl, Vogel, and Genscher. The Cold War has been an abstraction for them. Some essential lessons of the past which were self-evident for their elders are lost on them. They tend to take a more benign view of Soviet politics than the previous generation; in fact, they are not much interested in the subject and try to ignore it. Their attitude vis-à-vis the United States is more critical. As far as the postwar German leadership was concerned, America could do no wrong, whereas the formative years of the coming leadership were the years of Vietnam. Even though they have shed some of their former naive enthusiasm for countries such as Cuba, Vietnam, and other radical Third World nations, including Iran, and even though they accept in the abstract that they need—at least for the time being—the American nuclear umbrella, their enthusiasm for the alliance is limited, and there is an instinctive residue of anti-Americanism. These observations, needless to say, apply mainly to those sympathizing with the Social Democrats and the Greens, and it is also possible that further exposure to the harsh realities of international relations will further expedite a learning process. But not too much should be built on this prospect as far as the next few years are concerned.

The younger generation—those now in their late teens or early twenties—are less interested in ideology. In some respects, they resemble the "prerevolutionary" generation, for instance, in their concentration on career and personal life. But in many ways their political horizon is limited; even the elite among them tends to think in Central European rather than in global terms. Nor should one underrate the impact of indoctrination—by radio, television, and radical teachers belonging to the generation of 1967. Such indoctrination is more palpably felt in foreign than in domestic affairs; young people will be reluctant to accept uncritically propositions about domestic events and situations with which they are reasonably familiar. They are more easily influenced by information on faraway countries to which they have never traveled. Lastly, it ought to be stressed that the youngest generation is much freer than the previous ones of the burden of the past—meaning the heritage of Nazism. They find it unacceptable to behave like the previous generation, which felt it had to be on its best behavior in view of the crimes of the Third Reich. They believe that they cannot be made responsible for these distant events. A new patriotic, even nationalist mood can be discerned among this generation. Yet at the same time, they have to accept that, although they bear no responsibility for the crimes of a system which came to an inglorious end years before they were born, there is no escaping from history. They have to live with the fact that their country has been divided as the result of the war unleashed by the Third Reich.

The Future of West German Foreign Policy

But for a decisive shift in the global balance of power between the United States and the Soviet Union, and despite all uncertainties, German foreign policy in the years to come seems fairly predictable. Exposed strategically, and vulnerable economically, West Germany is subject to more constraints than almost any other European country (more so, in fact, than some smaller countries), and its freedom of action will be limited.

Predictability, reliability, and responsibility have been the preferences of German foreign policy in the past, and they will remain so in the future even though the old enthusiasm will no longer be there.

The Soviet "new thinking" in foreign policy and the growing independence shown by Eastern European countries has opened new vistas which did not exist until recently. The weakness of the East German regime has made the question of German reunification reappear high on the European (and world) agenda. There is growing enthusiasm to see Germany reunited in the next few years in both East and West Germany, and despite the considerable misgivings among Germany's neighbors, the old concept of two German nations and states as it existed between 1949 and 1989 is no longer valid and some new arrangement between the two Germanies will have to be found.

No one can say at the present time with any degree of assurance whether a confederation will gradually come into being or whether some other form of political coexistence will prevail. The question of immediate interest is how these new trends will affect the allegiance of the two Germanies to NATO and the Warsaw pact, respectively. The Soviet leadership has made it clear that it will not oppose Eastern European governments based on a coalition of Communists and non-Communists provided that these governments remain members of the Warsaw Pact. There have also been unofficial statements to the effect that in a long-range perspective the Warsaw Pact might be dissolved if this were matched by the dissolution of NATO. Such promises are bound to attract even more attentive listeners in West Germany than in other European countries. West German governments will stress that the danger from the East has substantially decreased, that nothing should be done now which might close the door to an eventual reunification, and that the West should make substantial concessions in negotiations with the East so as to hasten the process of peace. On the other hand, caution will probably not be altogether dropped in view of the uncertainties that will continue to prevail in the Soviet Union and Eastern Europe in the years to come. Present governments might be overthrown and present policies modified.

The differences between the two main German parties with regard to defense of the *Bundesrepublik* can be defined briefly as follows. The Kohl government believed until recently that present NATO strategy was by and large correct and indeed the only possible one. As a result of the INF Treaty, criticism of NATO strategy has been voiced, pointing to the increasing threat from the East as the result of Soviet superiority in conventional ground attack forces and of nuclear missiles with a range of less than 300 miles. Hence, the call for a second Montebello to follow up and update the decisions of the 1983 conference. This refers to the reduction and eventual abolition of short-range missile systems, to the modernization of other weapons systems, and to disarmament talks with the Soviet Union with the aim of removing the imbalance in conventional arms. With the extraordinary events occurring in Eastern Europe, this call may be even more important, albeit we can anticipate that the Kohl government's approach to the follow-up would be significantly altered.

The Social Democrats want a radical change in NATO strategy. They opposed the stationing of the missiles and want to replace the equilibrium concept by one based upon "conventional stability." Some have gone even further and in their blueprints aim at the withdrawal of U.S. forces from Germany in a midterm perspective, a reduction of the number of tanks and fighter planes, and the transformation of the *Bundeswehr* into some form of militia equipped with purely defensive weapons.

The weakness of this alternative concept derives from the fact that it is based on mutually exclusive premises—that the Soviet Union is much weaker than is believed in Washington and that, on the contrary, it is so strong that any future war would be fought on German territory and lead in a short time to total destruction.

A respectable case could perhaps be made for pacifism *tout court,* just as it could be argued that the likelihood of the use of nuclear weapons will decrease if conventional defense is strengthened. But the Social Democratic critics of present strategy want to remove nuclear weapons and to reduce the defense budget by cutting conventional defense. This mixture of contradictory elements is not serious; it is unlikely to become, in this half-baked form, official SPD policy. It ought to be recalled that Social Democrats opposed the "remilitarization" of West Germany under Schumacher, and that in the past—when the Social Democrats were in opposition—the idea to transform the *Bundeswehr* into a militia was for a while semiofficial party policy, which it is not at the present time. But these proposals came to nought, as the *Bundeswehr* was rearmed under SPD auspices. It cannot, of course, be argued that because the SPD government has ignored its pacifist wing in the past, this is bound to recur to the same extent in the future. It must be assumed that while the defense policy of a future Social Democratic government will not radically differ from the present one, it certainly will be different in emphasis.

All of these predictions are based on certain assumptions: that there will be no radical internal change in Germany; no breakdown in morale; no hankering after great-power dreams—that, in brief, West Germany will act more or less rationally and responsibly, having painfully experienced in her recent history the consequences of irrationality.

However, what if the far-reaching changes appearing in the international constellation, including the rejection of communist domination and economic decline in Eastern Europe and East Germany, are fully realized, opening in theory, or at least opportunities for German initiatives which do not at present exist? Or what if, alternatively, NATO should disintegrate or be reduced to a symbolic presence as a result of the changing European security environment? A decisive weakening of the Soviet military presence in Eastern Europe still seems unlikely during the next years, and there is no need to discuss it in this context. A weakening of NATO (such as, for instance, the withdrawal of the British Army on the Rhine for financial or political reasons, or the withdrawal of major U.S. forces) cannot be ruled out. To safeguard German sovereignty, a greater defense effort on West Germany's part would be needed in these conditions. Such an effort is certainly not beyond Germany's economic capacity, but it is equally certain that there will be many voices declaring it unnecessary because the outside threats have always been greatly exaggerated. Neither do the same circles favor closer European defense cooperation as an alternative to dependence on the United States. Their answer in the case of a breakdown of the Western Alliance would be neutralism. This is not the most likely contingency at present, but it cannot be entirely dismissed from speculation on the various scenarios of Europe's future.

The *Bundeswehr* in the Next Decade

The importance of the West German army for NATO is crucial. It provides 50 percent of the land forces and over 60 percent of the tanks in the central sector of Europe; it participates in the "Standing Naval Force Channel" and the "Allied Command Europe Mobile Force." West Germany is contributing almost as much as the United States (26.8 percent compared with 27.8 percent) to the NATO infrastructure program during the five years from 1985 to 1990.

The German *Bundeswehr,* employing 176,000 civilians on top of 495,000 soldiers, is one of the biggest employers in the country and, generally speaking, one of the biggest single economic factors. The German economy received from the

Bundeswehr during 1985 some $6 billion for goods produced and services rendered; orders provide employment for 250,000 workers and employers. This percentage is likely to remain more or less constant over the next three or four years and will act as a stabilizing factor in the German economy.

Modernization of the army during the next ten to fifteen years should proceed on the basis of decisions taken in 1984 and 1985. They provide, inter alia, for the following. There will be a slight increase in manpower from 1.27 million to 1.34 million beginning in 1987, on the basis of the *Wartime Host Nation Support* agreement. According to plan, the army will spend some $8–9 billion during the next ten years for new weapons systems, such as gradually replacing (after 1987) the *Leopard* I tank (introduced in 1965) with the *Panzerkampfwagen 2000* and the introduction of the multiple launcher MARKS and the Howitzer 155-1. In the German air force the *Phantoms* will be replaced by the European Fighter Aircraft ("fighter 90"), and the ground-to-air missile *Patriot* is replacing *Nike–Hercules*. The German navy is to introduce into service seven frigates and ten new-type minesweepers (343) and to modernize its submarines. However, in view of financial pressures, the more expensive weapons are likely to be affected. Some will be postponed, others cancelled altogether.

The German government has always been reluctant to accept the responsibility for an armaments industry; West German military research and development (R&D) is only about 7 percent of what the Pentagon spends for this purpose. At the same time, the German government wants to spend as much as possible of its arms expenditure inside Germany; it is at present at the level of 85 percent. A new R&D concept has been developed to make greater use of various coproduction schemes in fields in which German industry has a leading role. German strategy should rest as it has in the past on the principle of *Vorneverteidigung* (forward defense). Its capacity to do so will largely depend on the budgetary allocations for defense.

Altogether, German military force has been strengthened during the last three to four years, and, while at present a period of stagnation seems likely, the German armed forces are still the strongest in Europe in conventional terms.

To strengthen the efficacy of the armed forces and to prepare a possible alternative in case of the reduction of the American presence in Europe, the Bonn government has increasingly been looking for closer cooperation with France. There have been frequent meetings between senior commanders and defense officials from the two countries, and the collaboration reached a new height with the "Cheeky Sparrow" exercise in November 1987 and the planning for a combined French–German brigade. However, progress has been less than striking, partly as the result of relative French weakness in conventional strength. While both French leaders of the left and the right have expressed support for a close defense community, the French commitment so far has been ambiguous. Many Germans, including Helmut Schmidt, believe that the French forces would get instructions to participate in the defense of German territory only when it has become clear that the fight for Germany has already been lost. French leaders, on the other hand, have expressed disappointment that the Germans have shown insufficient appreciation for French moves away from Gaullism toward a partnership with Germany. Thus, the prospects for an increase in the importance of this bilateral relationship in the near future are not brilliant. Substantial advance can be achieved only at a time of crisis, when a more acute perception of threat prevails.

EUROPE'S NUCLEAR POWERS

France

Almost alone among major European countries, there is broad agreement among the main parties in France about national defense, and this is likely to remain the case in the years to come. For this reason, French defense is likely to be much less affected as a result of changes in government than British or West German. The two major problems that will increasingly confront French decision makers in the years ahead are, on the one hand, the character of French strategy and, on the other, the question of financial resources. There has been over a number of years a marked disparity between French military aspirations and the number of projects actually funded. While the economic situation has slightly improved with a recent rise in the GNP of 3.5 percent (between 1979 and 1987 it was merely 1.6 percent), a trend which is continuing, in the two years ahead, it is still not clear whether France will have the means and the willingness to allocate the resources needed for a minimum, credible, semi-independent defense effort. It has been said that de Gaulle gave France for a while the illusion of national defense on the cheap— "*Sanctuarization.*" But this has no longer been true during the last decade, and it will be even less true in the years to come.

The French dilemma is, in brief, as follows. The French nuclear forces rest on three pillars: its submarine force; surface-to-surface missiles stationed in the Albion plateau in Southern France; and the *Mirage* 4 bombers. The *Mirage* bombers will be phased out by 1996. As missile accuracy increases, the Albion silos are becoming increasingly vulnerable. A number of alternative solutions have been suggested, such as the SX missiles to be deployed on trucks (which would cost at the very least Fr 30 billion) and the production of neutron bombs, if no progress is made in disarmament talks by 1992. According to current plans, the range of the new *Hades* missile will be three times that of the *Pluton*, which has been in use for a long time. But its reach will still be only 350 kilometers and cannot, therefore, be considered a strategic weapon—in French parlance they are "prestrategic."

This leaves France's submarine force. Three submarines with forty-eight warheads are now on permanent patrol. The more modern *L'Inflexible* carries ninety-six warheads, and France intends to fit out four more submarines with the M-4 missile so that the *Force Oceanique Strategique* will have 288 nuclear warheads on permanent sea patrol.

However, many French strategists believe that this is not sufficient because antimissile defense is improving in the Soviet bloc. Some of them have suggested that five submarines with up to 500 warheads is the bare minimum to assure the force's credibility; others have insisted that up to fifteen submarines may be eventually needed.

The situation is further complicated by the fact that France has no effective satellite observation system, and French politicians have not been able to make up their minds whether to develop their own airborne warning and control system or whether to buy the U.S. system. The problem is in large part financial. The *loi de programmation* (1984–1988), the overall guidelines for French defense, underrated inflation and the rising costs of procurement; this has seriously affected both current defense expenditure and long-term planning. French spending on equipment now covers half of the defense budget and it is rising by 3.7 percent a year after inflation. Furthermore, the share of defense in the state budget declined from 17 to 15 percent during the first half of the 1980s. This had led to a postponement of important decisions—on the SX missile, on the building of a new fighter-bomber, and on two nuclear aircraft carriers. The production of the new *Leclerc* battle tank has been seriously delayed, which led to open protests by senior commanders (and their consequent dismissal). Conventional elements of the navy have been starved of resources, and by 1988 the land army was cut by 20,000 men. France has no large transport aircraft, seriously limiting the effectiveness of its *Force d'Action Rapide* (FAR)—for the recent deployment of French forces in Chad, U.S. aircraft were needed. Generally speaking, most available funds are at present going into submarines and the two nuclear aircraft carriers, and there is no serious planning beyond the year 2000.

The new nuclear submarines should enter service in 1994; according to official French sources, they will be relatively invulnerable in view of their quietness and the fact that they will operate from greater depths. These capabilities will make detection more difficult; however, this does not take account of possible progress in the technology of antisubmarine warfare.

The main weakness of the French defense effort resides in its imbalance and its lack of overall direction. Along these lines, there has been some disagreement between the major political parties. The Gaullists were aware of the fact that French deterrence may no longer be credible, but they offered no cures except to request a doubling or trebling of the submarine force. The Socialists want to put greater stress on tactical nuclear weapons (ANYT), and they have built up the FAR. Valery Giscard d'Estaing and his followers have advocated the neutron bomb but, aware of the fact that this would not constitute a final solution, favor closer cooperation with France's European allies, on the one hand, and America, on the other. Everyone is aware that greater budgetary allocations for defense will be needed, but there is great resistance against pushing the defense budget beyond 4 percent of the GNP (it is currently 3.7 percent). The idea of launching defense bonds has been raised, but again, no decision has been taken or is likely to be taken for some time.

The basic strategic questions which will face France in the years to come are those that have confronted it in the last decade. While a certain amount of movement seems likely, binding new commitments appear to be ruled out. This refers above all to France's relations with her European allies and the United States. There has been movement toward a more European character for French defense and, at the same time, somewhat closer links with the United States. The climate has improved, even though the old suspicions of integration ("*subjugation*") have not vanished. The crucial question remains. Where would France stand if West Germany or another European NATO ally were attacked? It has not been answered. On the one hand, there have been repeated and emphatic assertions that West Germany is for France far more than a mere *glacis,* a buffer zone, and on the part of the neo-Gaullists there have been occasional hints that the French umbrella could replace the American guarantee if Bonn so wanted. On the other hand, the neo-Gaullists have always made it clear that France could never give West Germany a formal guarantee that it would come to her assistance in the case of an attack. The Socialists, while in power, have also pointed out that it would depend on the decision of the president whether France would join the forward battle and that, as far as they are concerned, the strategic deterrent would not cover West Germany.

In recent years there has been closer French cooperation with Britain. Talks between the two countries have covered the reopening of French supply routes to British forces and combined troop maneuvers at some future date. More ambitiously, the possibility of collaboration in the nuclear field has been discussed so as to reduce wasted, duplicative efforts. However, because both the French left and right agree that nuclear deterrence rests on a *national* threat (a feeling shared by British leaders), it is unlikely that such collaboration will go beyond certain, well-defined projects in the near future.

Thus, France finds herself caught on the horns of a perpetual dilemma. There is the old fear that one day the United States may leave Europe and, following a dramatic change in the balance of power, Germany might become part of the Soviet zone of influence. In this event, France would be left in isolation with an unrealistic strategy of anti-cities retaliation. French decision makers dread a development of this kind, but they feel incapable of taking any steps that might forestall a process of this kind. To reassure West Germany, the *Force d'Action Rapide* was established. But it consists of altogether 40,000 men, and it has not escaped the attention of the Germans that the French forces in West Germany are deployed close to the French border so that they could be withdrawn quickly if need be.

Two schools of thought divide France's politicians and strategic thinkers. The neo-Gaullists stress the independent character of French defense, even though most of them have accepted that the idea of a sanctuary is no longer realistic. On the other hand, there are the "revisionists," who have accepted that an independent French defense posture is no longer feasible. The revisionists are sometimes accused of pro-Atlanticism by the neo-Gaullists, but this charge is justified only in part. For, while the revisionists do favor closer cooperation with NATO, "Euro-defense" is probably even closer to their hearts. They are certain that the old Gaullist belief in a separate role for France in European defense is outdated (if indeed it ever was realistic) and that, with rapidly rising procurement costs, France cannot afford any longer to cover the fees of membership in a superpower nuclear club for, in fact, it has difficulties in meeting the costs of any up-to-date, efficient conventional defense. The establishment of the FAR by the Socialists in 1983 was criticized at the time by the center-right because, in their view, it divided and weakened existing conventional forces. But when the rightists were in power they did not carry out substantial changes. This showed yet another time that the difference between left and right in France is one of emphasis rather than principle. True, the neo-Gaullists have traditionally preferred strategic over tactical nuclear weapons, but in the final analysis they advocated a mix of the two, and the same applies to the Socialists.

No mention has been made of French attitudes to SDI. The initial official position was that this initiative should be made unnecessary by disarmament talks. Later on, the view prevailed that technological progress could not be prevented because the superpowers would continue research in the fields associated with strategic defenses. In any case, France—whose strategy continues to rest on nuclear deterrence, with its forces targeted against major urban centers—also needed a project of this kind, to be undertaken with the help of her European allies. The French right-wing parties have shown less opposition against cooperation with the United States on SDI than the Socialists, but it is unlikely that any major initiative will be taken by anyone in the near future, short of a dramatic deterioration in the international climate. Generally speaking, the trend toward postponing important decisions, which has prevailed in French defense planning for more than a decade, is likely to continue. Some of the outstanding problems could be solved, temporarily at least, if an additional one percent of the GNP were allocated to national defense. But as the cost of modern weapons systems continues to skyrocket, even somewhat higher defense allocations would only postpone a solution by a decade. Questions of principle remain open, such as whether France would go to war only if her territory was immediately threatened.

While the defense of France in isolation from its European neighbors is now a mere delusion, the alternatives are either too costly or seem otherwise unpalatable. As a consequence, a basic change in orientation will occur only in the face of a clear and present danger. A slow, gradual retreat from Gaullism has taken place over the last decade and will continue during the next. The internal French consensus on defense seems to be considerably stronger than in Britain or West Germany. There are few illusions in France on the consequences of unilateral disarmament, and there seems to be more political will which may make it easier in the long run to overcome the present difficulties.[4]

France is now confronting and will continue to confront major domestic problems. The rise of the "National Front" in 1985–1988 points to a malaise in the body politic of the Fifth Republic. The electoral successes of the party in places such as Marseilles can be explained by reference to the influx of millions of foreigners, mainly from North Africa, and a failure in integration. But since the National Front also did well in districts such as Alsace–Lorraine, where there are relatively few new immigrants and where social conditions are by no means critical, the sources of the Le Pen phenomenon seem to be more complicated. Unemployment in France is high by any standards (9.5 percent in 1989), and there are fears that it may increase further despite various government job programs. While France has made some advances in high technology, it spends less on R&D than its neighbors, and it has considerably fewer scientists and technicians than West Germany and Britain. Despite these and other political, social, and economic weaknesses, French material and spiritual resources seem to be strong enough to prevent any major tremors that could decisively affect the military strength of the country.

The United Kingdom

Until a few years ago the debates on the future of British defense and the British contribution to NATO proceeded from the assumption that there will be a broad consensus between the major political parties in the future as there was in the past. But then, until the fall of 1989, the Labour party became unilateralist, committed to shutting down all U.S. nuclear facilities within several months of coming to power. Labour's policy was to scrap Britain's independent nuclear deterrent, *Polaris*, and its replacement, *Trident*, and destroy Britain's stocks of battlefield weapons. Currently, the entire NATO force of F-111 fighter-bombers and A-10 and antitank aircraft, as well as vital tanker and reconnaissance aircraft for use in conventional warfare, are based in Britain. The United States has eight major air bases and five naval sites in the United Kingdom; 30,000 U.S. military personnel are stationed there, about 27,000 of which serve in the Air Force. The Labour party wished to persuade the Soviet leaders to make reductions in strategic missiles in return for the decommissioning of *Polaris*. But since *Polaris* was becoming obsolete, the most that could be expected was that the Russians would get rid of an equivalent number of obsolete weapons systems.

The Labour party leadership at the same time insisted that the party would remain in NATO. Some leaders maintained that the United Kingdom would continue to make a substantial conventional effort to NATO; but this appeared doubtful, not so much in view of the opposition of the left wing, but as a result of economic and political difficulties.

Neither the United States nor Britain's European allies accepted the idea of a nonnuclear defensive strategy. British units would have had to be withdrawn from Germany and British officers removed from NATO command positions if the old Labour policy were carried out. Some analysts have argued that the reintroduction of conscription would make a nonnuclear strategy feasible. But it is most unlikely that any political party—least of all Labour—would adopt an unpopular measure of this kind. It is Labour policy to spend more for social welfare, whereas the reintroduction of conscription would mean an increase in the defense budget of up to 15 percent (cost of manpower, warstocks, equipment, etc.). Labour policy was based on a number of contradictions which cannot be bridged. It wanted to get the Americans out, but to remain in NATO; it wants a credible defense posture, but to spend less on defense; it wanted a more independent role for Europe, but no close European defense cooperation.

And in 1988, there were some indications of a possible shift in Labour defense attitudes. Neil Kinnock, the party leader, noted that his party's policy had to take account of the changes in international relations following the improvements in relations between the United States and the Soviet Union. Hence, his conviction that there was "no need now for a something-for-nothing unilateralism." And hence his desire to advocate a policy which would also bring about a significant reduction in the nuclear weapons of the Soviet Union—a viewpoint which was strongly criticized by the left wing of his party. Aware of the fact that Labour's prospects in the next elections would be small unless it scrapped its unilateral defense policy,

Neil Kinnock announced in the summer of 1989 that a future Labour government would no longer be bound by previous resolutions. In the fall of 1989, Labour dropped its unilateralist policy, accepting the new *Trident* submarines and the presence of U.S. nuclear weapons.

It cannot be ruled out that Labour will win the next general elections (1992). If they do, the Labour majority is likely to be small. In such circumstances, the Labour leaders may still insist on carrying out radical changes in defense policy; but in view of the opposition on the part of Washington and Britain's European allies, and their own professed desire to remain associated with NATO, talks with the allies would be initiated with the aim of reaching a compromise. In the meantime, defense expenditure would be cut and most or all development and modernization schemes frozen.

Another possible scenario is a Labour victory which would make it dependent on Liberal party support to attain a workable majority in Parliament. Because the Liberals had also gravitated towards unilateralism, it appears that agreement on defense between the two parties may not be impossible, provided that the more radical Labour demands have been watered down.

Lastly, there is the possibility of a fourth consecutive Conservative victory in the next general elections. In any case, substantial changes are now taking place in Britain's defense posture. With only slight exaggeration it can be said that the seven fat years for defense are now being followed by seven lean years. Britain's commitment to NATO's real growth target ended in March 1986. The defense budget in 1985–1986 was some $4.5 billion higher in real terms than it had been in 1978–1979. It was the longest period of sustained defense growth in thirty years, and Britain's record was better than that of most other NATO allies. However, the 1987 public expenditure white paper envisaged that over the next three years the defense budget would decline by about 6 percent in real terms.

The reasons for these planned cutbacks were both economic and political. North Sea oil revenues have declined from $16 billion in 1985–1986, and a further decline is expected. According to official British forecasts, the GNP was still likely to grow by 2–3 percent because of earnings from invisibles. These forecasts proved unduly pessimistic; actual growth in 1987–1988 was 4.6 percent. Nevertheless, continuous high unemployment and urban decay in the Midlands and Northern England, as well as the problems of Britain's competitiveness, continued to becloud the long-term prospects for the British economy. Substantial investments in these and other fields have become a political as well as an economic necessity. The question why Britain should spend per capita more on defense than its European NATO allies continues to be asked by the opposition with sometimes damaging effects. It has also been pointed out time and time again that NATO commitments, rather than the defense of the realm, absorb the bulk of Britain's defense resources.

Where could savings be made? A cut in spending on the Falkland garrison and the dismissal of part of the civilian work force in the United Kingdom would account only for a small part. *Trident,* which is to replace *Polaris* in the mid-1990s, is regarded by some as a wasteful luxury, but the present government commitment to this system seems unshaken. The alternative would be a cheaper deterrent based on cruise missiles. According to official estimates, believed to be too low by some, *Trident* costs will consume 7.5 percent of the total defense budget (and 15–20 percent of the procurement funds) in the peak years between 1988 and 1993. This is bound to cause a certain decline in Britain's conventional capacity, hence the opposition from inside the defense establishment. Other cuts could be made if the European fighter project were given up and if the British government were to buy its next fighter, as well as much other equipment, from America.

According to official British sources, there is no need at present for a fundamental review of defense policy priorities. It would be more accurate to say that there is an *urgent* need, but it might be pointless to carry it out except within the wider framework of a general NATO review. As defense spending will not increase, the alternative facing Britain is either to cut some of its NATO commitments or to cut down on the modernization of its equipment. One such possibility is the withdrawal of a quarter or a third of the British Army on the Rhine. These units may be, as some see it, an anachronism, but their transfer to Britain would certainly not save much money, whereas the political effect would be harmful. The cost of the British forces deployed in the FRG is less than $6 billion, but only a small part of this could be saved. At one time, cutting the surface navy was considered, which provides 70 percent of allied strength in the Eastern Atlantic. This idea was dropped following the Falklands war.

In 1985–1986, one nuclear-driven submarine entered service, along with an aircraft carrier, two destroyers, a frigate, and eight minesweepers. In 1986–1987, one submarine, one frigate, and three minesweepers followed. Four more submarines are on order, but, except for three minesweepers, no additional surface vessels have been ordered.

The British army is small, half the size of France's or Germany's; many of its tanks, guns, and half of its personnel carriers will be obsolete in a few years. All of which is to say that substantial savings could be achieved only by seriously weakening one component, or possibly two, of the British defense forces—by both weakening frontline strength and cutting down investment in new equipment for the future.

The better than expected performance of the British economy in 1987–1988 (and the Conservative victory in the general elections) made it possible to soften the shock foreseen in the 1987 Defense White Paper. The defense budget for 1988–1989 was some $800 million higher than provided for (£19.215 billion total). According to the new plans, it will be £20.560 billion in 1990–1991, but a downturn in the economy in 1989 makes it appear uncertain whether these goals will be met. Thus, both in absolute terms, as a percentage of the GNP, and per capita, the British defense budget continues to be one of the highest in Europe.

However, the budgetary pressures on UK defense continue, and equipment expenditure is now somewhat smaller than a few years ago. Among the equipment on order, the following should be mentioned: the *Trident* missile-carrying submarines, which have been since 1969 Britain's "strategic force of last resort"; four to eight Type-23 frigates; four *Sandown* minehunters; a third air defense missile regiment equipped with the *Starstreak* high-velocity missile system; SA-80 infantry weapons; one Boeing E-3 airborne early-warning aircraft; thirty-four more *Harrier* 6R-5 aircraft; and the *Boxer* 2 communication bearer system providing a nationwide secure and survivable communications system using fiber-optic technology.

The statement on the defense estimates for 1988 says the following on the personnel situation for the services:

> The manning position in the armed forces is broadly satisfactory, although there are shortfalls in certain areas. The Services face stiff competition from civilian employers, especially in those areas where there are national shortages of well qualified and able staff. Nevertheless, recruitment has reached 99 percent of the targets for both officers and other ranks; shortfalls in officer recruitment have been in specialist areas, for example medical officers.
>
> Competition with the private sector is likely to increase over the next few years as demographic trends tend to fewer young people being available for employment, while the expansion of the economy increases the number of jobs available. To help forestall these difficulties the Services are seeking to retain trained manpower for longer periods. This has the added benefit of ensuring a better return on our investment in training those retained, of reducing pressures on training organizations and of making more trained and experienced manpower available for front line tasks.[5]

Given the polarization of public opinion in Britain, the future size and direction of the British defense effort is more difficult to predict than that of any other major country. Anti-American influence on the left has increased in strength over the last decade, but a Labour party dominated by extremist trends is unlikely to win a general election, a fact which Kinnock has recognized. While extra-parliamentary political action is possible against a background of high unemployment, urban decay, and ethnic conflict, it is unlikely that the disruptive elements will prevail over the forces of law and order. On the contrary, an outbreak of violence on a wide scale would strengthen the right wing. At a time of internal danger or serious economic breakdown, the moderate forces of both right and left would almost certainly draw closer together, as they did in the 1950s and 1960s; however, in any but an extreme situation such collaboration is unlikely.

Radical change is still unlikely in the near future as far as British defense efforts are concerned. But decisions of far-reaching consequence may be taken by default if, for instance, funds were withheld from the development of weapon systems as the result of a budgetary squeeze. In these circumstances, Britain would no longer have a serious nuclear deterrent after the mid-1990s. Alternatively, the navy, the air force, or the army could be weakened to an extent that they could no longer fulfill the function of defending the homeland, let alone take part in the forward defense of the European mainland. The navy could be reduced in strength, but in this case it would no longer be capable of landing troops in wartime. The natural instinct of British governments has been to inflict equal misery, that is, to reduce allocations to the air force (for instance, by cancelling the replacement of *Jaguar*), the army (by delaying or cancelling procurement of the new battle tank), and the navy (by paring off a third of its aircraft carriers).

This leaves the question of the British nuclear deterrent. The political and military alternatives, ranging from relying on the United Nations for the defense of legitimate British interests to planting more trees in Germany (as has been suggested) so as to slow down a Russian advance, are not serious. Antinuclear feeling is deeply rooted in the United Kingdom, and despite the recent changes in the Labour party's position on unilateral disarmament, Britain under a Labour government could still become the focal point of a major NATO crisis at some future date. Given Britain's lack of enthusiasm with regard to truly close cooperation with Europe, it is unlikely that such a crisis would be productive—that is to say, that it might lead to a European alternative to NATO. While the longing for a nonnuclear defense in West Germany is equally strong, the absence of a common border with the Eastern bloc tends to breed in the United Kingdom illusions about the safety of the island as the result of unilateral disarmament and the possibility of neutral status—illusions which West Germans can nurture only with a considerable intellectual effort. No one can say for certain how far opposition parties will go in carrying out their programs once they are in power. But it may be no accident that the program of the British Labour party has gone further in its demands to scale down Western defense than the program of the German Social Democrats. Whatever party (or coalition of parties) will be in power in the years to come, there will be no significant progress in British defense. Under the Conservatives it has leveled off (to use the official term); under Labour it would substantially decline.

THE BENELUX COUNTRIES AND SCANDINAVIA

To understand present attitudes toward defense, as well as their likely future development in Scandinavia and the Benelux countries, it is of crucial importance to remember that these countries have been neutral and neutralist for most of their modern history. Only as the result of the Second World War and the German occupation of all these countries (excepting only Sweden) was there a willingness to give up neutrality after 1945 and to join NATO. Since then four decades have passed, new generations have come to the fore, and the awareness of the vulnerability of these small and exposed countries has very much receded. This concerns not only attitudes toward NATO; the willingness of these countries to undertake a serious effort for national defense has also declined. The strength of the Belgian army in the 1950s was 158,000; it had declined to 90,000 in 1987. Dutch defense spending in 1962 was 19 percent of the state budget; it decreased to 9 percent in 1984. Even Sweden, proud of its traditional neutrality and willing to make sacrifices to maintain it, is aware that its defense effort is no longer sufficient in the regional context. Only Norway has undertaken during the last decade a serious effort to keep and improve its defense readiness.

Some observers have explained the erosion of the national will in the "Scandilux" countries with reference to a malaise expressing itself not only by way of aversion to NATO or anti-Americanism, but as a general tendency toward isolation and withdrawal. The enthusiasm for more unity in Europe has also notably cooled; Denmark, once one of the most enthusiastic supporters of European unity, is now one of the most recalcitrant. The general discontent has expressed itself in a breakdown of domestic peace almost unprecedented in the history of these countries—mass strikes in Denmark, Norway, Finland, and Iceland; growing difficulties with the functioning of the welfare state system in Sweden; and frequent government changes in Belgium and Holland in the 1970s and early 1980s. Yet others have pointed to the fact that more radical leaders have come to the fore in the Scandilux socialist parties. This is particularly true with regard to the Dutch and Belgian (Flemish) party. This new leadership is by no means Marxist in inspiration, but pragmatic and opportunist. The lack of realism in foreign affairs and defense merely reflects widespread popular feeling—the unwillingness to shoulder responsibility, the preoccupation with high living standards, and the tacit assumption that the stronger powers will somehow take care of the security of the small nations.

These negative trends have been most pronounced in the case of Denmark. The unwillingness to increase defense spending and to enter new defense commitments is usually explained with reference to the economic downturn during the late 1970s. However, in Denmark a nationwide debate on whether to leave NATO took place as far back as 1969, and Danish antidefense trends also predate the economic difficulties and the rise of unemployment in that country. The number of Danes on active military duty has sunk from 50,000 in 1966 to about 30,000 at the present time, of which 30 percent are conscripts. Negative demographic trends can explain this trend only in part. The percentage of defense spending in terms of the GNP has declined during the same period from about 2.8 percent to 2 percent—that is, by more than a quarter. Popular attitudes toward NATO and defense in general are contradictory. A great majority of the population support the defense forces, and 64 percent still believe that a Danish defense effort helps to ensure peace. But only a third (or less) believe that the Danish defense forces can be effective, and only 50 percent are willing to fight if need be. NATO membership is supported by 70 percent of the Danes, but the same percentage oppose nuclear defense.

The Social Democrats and the Radicals believe that domestic social spending is far more important than defense expenditure. Only the most extreme groups in Denmark, as in the other Scandilux countries, want to leave NATO, and support for alternative neutralist schemes, usually put forward by Finland, is strictly limited. But the weakness of successive governments, created in part by a surfeit of political parties, has made a consistent defense policy virtually impossible. Because Denmark does not produce arms, most of the funds allocated to weapons acquisition have to be spent abroad; this necessity further contributes to the lack of support for defense. The Danish malaise, or "Denmarkization," can be explained in part with reference to the general crisis of the smaller European countries. In an alliance of unequals, the junior partners tend to contribute less in relative terms to the common endeavor than the senior partners. But this does not explain why the Norwegian armed forces rose in numbers during the period that the Danish armed forces decreased, and that they are now larger than those of Denmark despite the fact that Denmark has a greater population. Nor does this explain why Belgian per capita defense spending went up, while in Denmark, it went down, and this despite the fact that Belgian economic difficulties were (and are) graver than those of the Danish.

The Danish defense crisis reached a climax in 1988, when a conservative-led coalition was brought down by a left-sponsored resolution demanding an assurance from visiting NATO warships that they were not carrying nuclear weapons. The subsequent elections helped to patch up the status quo, but in the view of most observers the current situation is unlikely to persist.

While the Danish Social Democrats had supported the double-track decision of 1979, their stand changed radically when in opposition. The Danish left came to support new policies of "defensive defense," meaning the abandonment of NATO forward defense strategies, including the withdrawal of one of the three Danish brigades from the joint Danish–German command defending North Germany, and also a switch from forward naval defense in the Baltic to coastal defense, the scrapping of big naval vessels, and the disbanding of large military units.

Critics of the defensive defense strategy pointed out that these new proposals not only ignored military realities, such as the fact that each year many Soviet vessels cross Danish territorial waters, including at least some equipped with nuclear weapons, but that the new nonprovocative strategy is directed also against conventional defense. According to polls taken in 1982, 82 percent of all Danes favored a nuclear-free Scandinavia, a concept first advocated in the late 1950s by Finland in coordination with the Soviet leadership (Soviet territory, including the Kola peninsula and the Leningrad military district, was not to be included). Yet, at the same time, Denmark continued to rely on Western support in the case of an outside attack. When Britain, in retaliation for Danish foot-dragging, announced in 1987 that it was considering reducing the number of troops available for the defense of Denmark and Baltic approaches in time of war, there was considerable indignation in Copenhagen.

Denmarkization emerged during the 1970s as a new term in the political dictionary, as did "Hollanditis." There were no more ardent Atlanticists and better Europeans in the first two postwar decades than the Dutch. But around 1970 the defense consensus began to crumble, and Holland entered a negative phase in its development which manifested itself in many respects; a certain recovery in some fields can be detected in recent years, as shown by the results of the general elections of 1986. The issue at stake was not only (and not mainly, as often believed) opposition to nuclear arms and the stationing of new weapons systems. It was rather a general crisis of Dutch society and establishment values, the turning of the churches against defense, and the foreign political reorientation of the Dutch Labor party (PvDA), like many West and North European Social Democratic parties, to a quasi-neutralist position. The Dutch demonstrations against INF deployment were among the largest in Europe at the time; even the center-right coalition showed reluctance to push through the NATO decision.

The essential facts about Dutch defense are to a certain extent contradictory and can be interpreted in different ways. While per capita defense spending was considerably higher in the 1960s than now, there has been no further decline since 1970, and real defense expenditures have continued to grow. The number of Dutch soldiers has decreased from 162,000 in 1962 to 108,000 in 1988, but no further reductions are planned for the next decade. Dutch leaders have claimed that their NATO partners have focused their attention on quantitative reduction in Holland rather than qualitative improvements. Dutch recruits are said to have performed tolerably well despite the fact that they were permitted to wear beards and, generally speaking, were treated by and large like civilians. There has been a certain amount of modernization in the Dutch navy—traditionally a strong pillar of the defense of the Netherlands.

As suggested, Norway has been by and large a notable exception to these trends in defense. Opinion polls have shown greater support for defense spending in 1982 than in 1971, and support for NATO has remained high. Danish defense spending was $1.8 billion in 1987, whereas Norwegian defense spending was $2.7 billion—despite the fact that the GNPs are about equal. The national consensus with regard to defense has not been broken, and this consensus also encompasses the Labor party, which reaffirmed its support for NATO in 1981, supported a 3 percent increase in defense spending in 1984, and has also accepted the prepositioning measures for a U.S. brigade. (Norwegian membership in NATO has rested from the beginning on the principle of nonbasing in times of peace.)

The Norwegians' awareness of the dangers to which their country is exposed is, no doubt, connected in part with the fact that alone among the Scandilux countries which belong to NATO it has a common frontier with the Soviet Union. On various occasions it has come under military and political pressure from the East; the Soviet presence in Svalbard (Spitzbergen) and the deployment of Soviet naval forces in the proximity of the Norwegian coast are constant reminders. The main problem that will face Norway is a dearth of manpower as a result of the declining birthrate. To remedy this, the call-up of young women in the 1990s has been discussed. Other major problems are the spiralling costs of modern technology, which will affect above all the air force. On the other hand, the Norwegian fleet (with fourteen submarines) has been traditionally strong and relatively well funded.

In contrast to Norway, Belgium has not adhered to the NATO decision to increase its defense budget by 3 percent annually. There was insufficient political support for such a measure. When the Supreme Commander of NATO suggested that Belgium should make certain cuts in social spending, Belgian socialists argued that before the NATO resolution Belgian defense spending had risen for eight consecutive years, whereas American defense spending had decreased during this period. This kind of argument, while true, was hardly pertinent.

An interesting and specifically Belgian development took place inside the Socialist party. Whereas in past decades the French-speaking (Walloon) element had been traditionally more radical and pacifist, this has changed in recent years. The Francophone Socialists have taken their cue to a considerable extent from their French comrades, whose attitude toward the Soviet Union has been critical and who have been on the whole in favor of a substantial defense effort.

The Flemish Socialists, on the other hand, whose influence inside the party has grown, are more orientated toward the Dutch, as well as the West German and Scandinavian, Social Democrats. It was among the Flemish that antinuclear moods have found in recent years the most ardent advocates. Belgium has been subject to serious economic difficulties since the early 1970s, mainly as the result of the failure to modernize the traditional industries in the south of the country. This led to substantial unemployment, and, as a result, economic and social questions have figured topmost on the national agenda. There is less doubting the need for strong defense in Belgium than in Holland and Denmark, and the commitment to NATO continues to be fairly strong. But the bad economic situation has affected the operational capacity of the Belgian armed forces and also their modernization. The results will be felt for years to come.

With all the political, economic, and social difficulties that have been mentioned, no sudden changes should be expected in the Scandilux countries. Voters in these countries have been conservative in their habits. To the extent that shifts in policy have taken place, they have more to do with generational change than ideological factors. The peace movements are strong, but they seem to have reached the height of their influence in the 1970s and early 1980s. There is a strong desire in all of these countries to maintain their independence, and because they all are situated so much nearer to the Soviet Union than to the United States, self-interest dictates that the Western Alliance should remain in force so as to counterbalance the threat from the East. What if Soviet pressure should grow? Would this lead to a collapse of the alliance or, on the contrary, to its strengthening? One cannot possibly provide an answer that will cover all Scandilux countries. The trends in favor of appeasement are strong, as they were on the eve of the Second World War, but the prospects are nevertheless good that most of the nations concerned would increase their defense efforts. There is, of course, the danger that defense will have been neglected to such an extent that any attempt to catch up may appear hopeless.

In the final analysis, Scandilux attitudes toward defense and NATO will depend in the remaining years of the century not so much on what America will be doing (or will refrain from doing), but on Soviet foreign policy in Europe and in other parts of the world. The Scandinavians and the Benelux countries certainly want to believe that there is no real threat from the East, a belief which has been reinforced by changes in the Soviet Union and Eastern Europe, the improvement of relations between the superpowers, and the dispatch of high-level Soviet emissaries in 1987–1988 to Scandinavia, impressing on their hosts the peaceful intentions of the Gorbachev Politburo. These missions had a certain impact on politicians and parties. Those in charge of defense have been less impressed in view of the existing imbalance of forces. As the "Military Balance 1986/7 in Northern Europe," published in Oslo, says:

> The Soviet Union and the other Warsaw pact members have considerable naval forces in the Baltic. The composition of the Soviet Baltic fleet seems to indicate that its primary task is to achieve control of the Baltic and carry out operations to open its exits. . . . It is not known for certain whether the Soviet forces on Kola have an offensive task, but the Soviet military doctrines are clearly offensive. The organization and equipment of the Soviet armed forces and the presence of air mobile units, naval infantry and amphibious capability in the area also indicate offensive intentions.[6]

All this does not mean that Gorbachev's emissaries have been insincere with regard to their professions of peace. However, it does mean that, for whatever reasons, even under *glasnost* and *perestroika* the buildup of Soviet forces has continued more quickly than in the democratic countries of Western and Northern Europe and that, as long as intentions might change, the issue of capabilities remains a matter of great seriousness.

NATO'S SOUTHERN FLANK

Italy

The basic facts concerning Italian defense are few and relatively uncomplicated, at least in comparison with other Western European countries. Italy's per capita defense spending has gone up slightly in the 1980s (from 2.4 to 3 percent of the GNP), and the total number of Italian soldiers, sailors, and airmen is close to 400,000. However, the quality of their equipment and their training is not comparable to that of West Germany, the United Kingdom, or France. A ten-year restructuring and modernization plan decided upon in 1975 was delayed and will be completed only in 1991. In contrast to other European countries, there has been no striking increase in antidefense or anti-NATO feeling; the decision to accept 112 cruise missiles did not encounter much resistance. On the other hand, Italians have felt less threatened than most other European nations. An invasion from the South or a Soviet naval attack has been thought unlikely. Relations with Yugoslavia, which were bad for a long time, have greatly improved. This leaves the "Goriza gap" in the North, the most likely land route for an invasion, on which the attention of Italian military planners has been focused for many years. It also leaves the possibility of certain non-NATO crises, for instance, as the result of an attack carried out by Muammar Qaddafi.

In recent years there has been a certain shift in Italian military thought as the result of the weakening of the Allied naval presence in the Mediterranean (above all the British, but to a certain extent also the American) and the growth in strength of Soviet naval forces in the region, as well as concern about certain Arab countries allied with the Soviet Union. Italian military thinkers have made various proposals on structuring their armed forces to achieve a balance between NATO and national requirements. They found their expression in part 3 of the Italian White Book of 1985, with its discussion of the "*Modello di difesa*" which, broadly speaking, still remains in force.[7] However, these proposals have not led to any fundamental reorientation nor to any major effort to strengthen Italian forces; there is a notable absence of any feeling of urgency, which will be reinforced by the reduced perception of the Soviet threat resulting from changes in the Soviet Union and Eastern Europe. Defense effectiveness has been somewhat affected by inflation and other economic difficulties and there has been little support for a major new defense effort. Instead, schemes such as a "total defense" model and "regionalization" have been discussed; total defense refers to something akin to guerrilla warfare in the case of an invasion, and regionalization refers to the dispersal of the army all over Italy to help the civilian authorities in cases of disaster.

These ideas clearly were not thought through and were eventually dropped. It has been realized that the Italian navy needs more destroyers and frigates, the air force needs more aircraft (the Eurofighter), and the army needs strengthening. Military transport and antitank and antiaircraft capabilities need improving. The Italian defense effort is hampered by rivalries among the army, navy, and air force; by the absence of an overall binding strategic concept; and by certain outdated bureaucratic regulations. Having commissioned a new aircraft carrier, the *Giuseppe Garibaldi,* the Italian government suddenly realized that according to one of Mussolini's laws, which had never been abrogated, Italy was forbidden to equip its ships with planes, because, in the words of the *Duce,* "all of Italy was a giant aircraft carrier."

Political and economic stability in Italy is considerably greater and is likely to remain greater than many outside observers believe, confused by frequent changes of government. Despite the presence of a major Communist party, the basic consensus on defense is likely to remain stronger than in Britain and West Germany, which have no strong Communist parties. On the other hand, social and economic needs will take first place in the public consciousness. Since NATO's southern flank has traditionally been attributed less importance than the central front, Italy has found it easy to take a low profile in NATO. This is unlikely to change in the years to come.

At the same time, Italian governments have proved more reliable NATO partners than other South European countries. When the Spanish government demanded the withdrawal of seventy-two F-16 jets (the nuclear-capable fighter-bombers of the 401st Tactical Fighter Wing) from Torrejon, Italy expressed readiness to provide bases for this force. While Italy, like Germany, has pressed for quicker progress in the negotiations concerning chemical weapons, it has not opposed modernization in the nuclear field. This is based on the belief that deterrence has to be up to date. Italian defense officials have stressed that European defense should be strengthened so that the Soviet Union will no longer have the capacity to engage in a surprise attack. Hence, Italy is willing to cooperate in bilateral and multilateral efforts together with her allies, in NATO and the Western European Union, on the land, in the air, and on the sea (NATO On Call Force Mediterranean). While Italian governments have warmly welcomed Gorbachev's foreign political initiatives, they have shown greater circumspection than most other European governments in accepting them at face value.

Spain

NATO interest in Spain has been limited, by and large, to air and naval bases; the NATO planners have also been concerned with the idea of an Iberian redoubt if conventional defense against a Soviet invasion north of the Pyrenees should fail. For many years under the Franco regime, the Spanish armed forces, while impressive as far as figures are concerned, fulfilled functions more suitable for a police force. They were ill equipped, badly trained, their prestige was low, and there were too many senior officers (many of whom spent much of their time in other jobs).[8]

In recent years there have been reforms and improvements based on the *Joint Strategic Plan* and *META,* the plan for the modernization of the land army. The Spanish economy has stagnated in recent years, and so has defense spending (2.3–

2.5 percent of the GNP). Progress as far as modernization is concerned has been slow and modest, and for years to come no change should be expected in this respect. Only the elite units of the Spanish army are considered as being up to NATO standards. The state of the Spanish navy is better, its prospects are brighter, and it has received a greater share of defense spending. A second aircraft carrier entered service in 1988, four frigates were added to the eight that exist, and two more submarines were added to the eight that are in service. The new vessels are equipped with suitable missiles, and their antisubmarine warfare capabilities have been improved. This task force will operate in the Mediterranean. At the same time, patrol aircraft and minesweepers are entering service.

It is generally thought that in an emergency the Spanish navy—in contrast to the army—could be integrated almost immediately into NATO. The Spanish air force is small—less than half the size of Iraq's, by comparison. Its share of the defense budget has gone up since the days of Franco from 20 to about 30 percent. But the most modern aircraft at the disposal of MACOM and MATAC (the Air Combat Command and the Tactical Command) are *Mirage* F-1s. Only recently have F-18s been delivered to replace the more obsolete aircraft.

National votes on Spain's adherence to NATO showed a notable lack of popular support for Western defense. The proposition was eventually adopted, but only after a considerable effort by the ruling Socialist leadership, which invoked European solidarity—more popular in Spain than the NATO link—and after a great many conditions had been attached to Spanish membership in NATO. Such reluctance should be explained partly with reference to anti-Americanism, but equally to historical and psychological factors. Spain has been neutral for a long time in her history, including during both World Wars, and largely isolated from the rest of the world. Franco managed to keep out of Hitler's alliance, and there is the temptation not to get involved in entangling alliances at the present time. Others advocate a Spanish orientation to the South, that is, the Mediterranean and North Africa. These visions are not rooted in reality. The North African countries are not particularly well disposed toward Spain nor are they important trade partners. It could well be that the Mediterranean orientation stems from a feeling of inferiority vis-à-vis Europe rather than genuine links with North Africa, even though the speed of the Europeanization of Spain during the last two decades has been remarkable.

Contrary to dire predictions and an uneven economic showing, there has been remarkable stability in Spanish politics during the last decade. The domestic record of the Socialists has not been impressive, but the center-right has been divided, and its performance in the given circumstances might not have been better. Far-reaching concessions have been made to the "minorities," and even the Basque *Euzkadi Ta Azkatasuna* terrorist organization is no longer a threat to the existence of the political regime. But for the worrying unemployment figures (17 percent), one's prediction for the next decade in Spain would be in a more positive vein. But even if there will be no improvement in this respect, a dramatic radicalization toward left or right is unlikely, as Spain seems to be immunized, at least for some time to come, against nondemocratic trends. What probably will happen is that Spain will be looking inward, giving far higher priority to its internal problems than to foreign affairs and defense.

Following the removal of the U.S. Torrejon airbase in partial fulfillment of the Spanish Socialist's electoral promise of 1982 (that *all* U.S. bases would be removed), Spain expressed interest in closer European defense projects. Spain has also suggested that it should accept the responsibility for a sea area stretching from the Balearic Islands in the Mediterranean through Gibraltar to the Canaries. However, it is generally believed that such an undertaking exceeds Spanish capabilities. There are no demands at present to remove the other U.S. bases (Rota, Saragossa, and Moron), and it is unlikely that these bases will figure prominently in negotiations with Spain's allies in the years to come. Spain has also suggested the opening of its territory to supplies and personnel in wartime. One light airportable brigade and the Spanish foreign legion have been offered to NATO for use in an emergency in Northern Italy. However, NATO commanders are not likely to show much interest in these proposals.

The Eastern Mediterranean

While the military situation in the Eastern Mediterranean—which involves two NATO members, namely, Greece and Turkey—is outside the purview of this study, a word or two may be in order. Greece's NATO membership may appear to be an anachronism considering the neutralist policy of the previous government. But concern over national security has been paramount in the Greek decision to remain within NATO. It is likely that relations with Turkey would be far worse than they are today if Greece left NATO. Even though it is difficult to believe, the two NATO members might go to war with each other if Greece withdrew from NATO.

Greece's armed forces are 80 percent American equipped and, in a period of economic difficulties, U.S. military aid is more than welcome from Athens' point of view. While there has been an improvement in relations with Turkey since 1988, the Greek government still believes that as long as the country remains within NATO it has a better chance to press for an American commitment to remove the Turkish army from Cyprus. These factors have made it difficult for Greece to insist on the removal of the four major U.S. bases in Greece—the Hellenikon air base, a naval communication center on the mainland, a naval support center, and the Iraklion air base. Although a former Greek prime minister pledged to remove *all* bases, no deadline was set and no action was taken to remove them. The demands for removal were concentrated on those facilities in thickly populated areas where American personnel were most in appearance.

This policy is unlikely to change in the near future, and the same is true with regard to Turkish attitudes toward cooperation with NATO. Turkey has been more positive than Greece, partly, no doubt, in view of the common border with the Soviet Union, but also because of the urgent need for the modernization of Turkey's armed forces. Turkey's demands vis-à-vis the United States do not concern the removal of bases but the supply of more economic aid and the increase of military supplies. While there are anti-American forces inside Turkey, and while the increasing influence of radical Islamic views could have a certain, limited impact on Turkish foreign policy, it is unlikely that radical change will take place.

SOVIET FOREIGN POLICY AND THE FUTURE OF NATO

Soviet foreign policy under Gorbachev has been greatly modified but, by and large, these changes have been less dramatic than in domestic affairs, and in Soviet military policy the changes largely belong still to the future. True, civilian and military strategists have clashed about the concept of *dostatochnost* (sufficiency), recalling similar debates in the West. Soviet experts have admitted that mistakes have been committed in the past, such as engaging in an arms race with the United States which proved to be ruinous. Even the crushing of the Prague spring reform movement in 1968 has now been denounced by the Soviet leadership. There has been a demand to redefine the national interest of the Soviet Union, and there have been calls against Soviet adventurism in the Third World which affects the overall image of the Soviet Union as a peace-loving country. Such "new thinking" has already had some impact on the conduct of Soviet foreign policy, as shown by the withdrawal from Afghanistan, the greater readiness to make compromises in arms talks, and the announcement of unilateral cuts in conventional forces by President Gorbachev, in his U.N. speech in December 1988. However, it is too early to predict how far the new thinking will go. Drastic reductions in the allocations for the Soviet armed forces have been suggested, but *Red Star* has pointedly recalled that one of the reasons for Nikita Khrushchev's fall was his arbitrary policy of starving the Soviet armed forces. Nor is it clear to what extent Soviet interests in Europe are likely to change. These interests, by and large, predate the revolution of 1917; some rulers of Russia pursued them with greater emphasis than others, but no one since Peter I has ever given them up. If West European governments will nevertheless be influenced in their security decisions, this has less to do with fundamental changes that have taken place, or are likely to take place, in Soviet policy and Soviet and Eastern European politics than with the well-known and perhaps intractable inability of democracies to engage in a reasonable defense effort except in the presence of a clear and immediate danger. Another argument that will be increasingly adduced is the internal weakness, real or imaginary, of the Soviet Union, which will not be overcome for years. Facing so many economic, social, and political problems at home, not to mention the nationalities question, can it not be taken for granted that the Soviet leadership will be fully preoccupied with domestic affairs and that as a result pressure will not be exerted vis-à-vis its European neighbors? This assumption is correct as far as it goes; domestic affairs will certainly take precedence over foreign policy. But, while domestic affairs have almost always taken priority in Soviet history, this has not prevented the Soviet Union from engaging in major and costly military buildups.

The immediate threat has receded in the perception of all European governments, especially with the extraordinary events taking place in Eastern Europe, and this will no doubt have major consequences with regard to defense spending and, generally speaking, to collaboration in the military field—more among the smaller European nations than among the larger, but ultimately it may affect all of them. The Atlantic Alliance came into being as the result of the perception of a common threat. If the feeling of the acuteness of the threat recedes or disappears altogether, the alliance will lose its raison d'être, and its fate will be that of other alliances in history when they outlived their purpose. This point has clearly not been reached yet, but questions arise as to what extent perceptions in Europe (and partly also in Canada and the United States) have moved in this direction. This is likely to be the crucial issue concerning the defense of Europe in the years to come.

NOTES AND REFERENCES

1. *The Financial Times*, October 31, 1989, Sec. I, p. 3.
2. According to projections by the *Bundeswehr* general staff, the German army will face a deficit of DM 26 billion in the 1990s, of which half is needed for material incentives to attract soldiers and officers to sign up for long periods of service. This sum does not include the means needed for modernization of equipment to retain the capacity for forward defense, which now exists, but which will be gradually lost unless the brigades equipped with old tanks (*Leopard* I) will be supplied with *Leopard* IIs and up-to-date arms systems. This would involve an additional DM 11 billion for which provision has not been made.
3. West German defense spending reached a low in the late 1970s with DM 31 billion in 1975. It rose to DM 50.8 billion in 1987.
4. There is a strange and not altogether explicable dichotomy between French official policies (and the position of the major parties) and the results of public opinion polls, which have fairly consistently indicated stronger pacifist, even defeatist, trends than in other major European countries. There appears to be a marked contrast between national pride and the willingness to make sacrifices for this purpose. But such discrepancies can be found, albeit less markedly, in other countries as well. For all one knows, public opinion polls in the Napoleonic age would also have indicated a nonbellicose mood. While there may be grumbling about defense spending, a strong leadership will still have overwhelming public support for expenses that are perceived to be necessary.
5. *Statement on the Defence Estimates 1988,* vol. 1 (London: Her Majesty's Stationery Office, 1988), pp. 51–52.
6. The Norwegian Atlantic Committee, "The Military Balance 1986/7 in Northern Europe" (Oslo: n.d.), pp. 2–6.
7. Ministero della difesa, *Libro Bianco,* 1985, pp. 35–60.
8. The Spanish army strength was cut from 300,000 to 230,000 between 1983 and 1988, but most experts believe further cuts might be beneficial.

2

Western European Defense

The Next Ten Years

INTRODUCTION

It is still quite likely that by the year 2000 the security of Western Europe will rest on the same three pillars as today: the alliance with the United States to provide the indispensable counterweight to the Soviet Union and its inherent geostrategic superiority; the combination of deterrence and defense with détente and arms control, as enunciated in the "Harmel Report" of 1967; and a strategy of "flexible response" to Soviet aggression, linking conventional, theater nuclear, and U.S. strategic nuclear forces for purposes of both deterrence and, should deterrence fail, defense. However, it is also virtually certain that each pillar will increasingly be subjected to stresses which have their roots in NATO's built-in contradiction between U.S. and Western European interests regarding nuclear escalation and in political, generational, and economic trends. By 2000, then, continuity in the structure of Western Europe's defense will probably have prevailed over substantial change more for lack of a feasible alternative to, than because of any intrinsic strength in, NATO's fiber.

The alliance with the United States, indispensable though it is to both sides, will decrease in political saliency as well as strategic importance owing to increasingly inward-looking successor generations on both sides of the Atlantic, whose outlooks are drifting apart; the new era of détente in Europe, which will find Western Europe's attention riveted on cooperation with the East and America's strategic focus shifting, once again, to "out-of-area" contingencies; European reluctance to jeopardize détente through direct involvement in out-of-area confrontations; and a growing U.S. inclination, stemming from the aforementioned trends, to unlink her strategic fate from Europe's—currently represented by the Strategic Defense Initiative (SDI), the Intermediate-Range Nuclear Forces (INF) Treaty, the talk about "regionalization" of deterrence, and renewed calls for the reduction of U.S. troops in Europe. The relative economic decline of the United States vis-à-vis

Western Europe and Japan will reinforce this last trend. Moreover, it is highly unlikely that any form of Western European defense cooperation—let alone integration—over the next ten years will be able to substitute for the vigor lost in the transatlantic partnership.

One may safely predict that this further loosening of transatlantic ties without compensating progress in Western European defense integration will lead to an increase in neutralism, especially in West Germany. With the dramatic changes occurring within the Eastern bloc, still risky alternatives such as denuclearization, "common security" with the East, and, ultimately, a collective security system for Europe from the Atlantic to the Urals will inspire increasing sections of the German political elites, including members of the Christian Democratic Party of Germany (CDU),[1] and the present security structure will be regarded as facilitating a safe transition to these future forms of defense rather than as an end in itself. Only two developments could arrest this trend in attitudes: another sudden deterioration in East–West relations, perhaps following progressively worsening turmoil inside the Soviet empire; or a French decision to establish a genuine security partnership with the Federal Republic of Germany (FRG) within NATO in the form of a commitment "to defend [her] with *all* military means."[2] This decision would extend French nuclear deterrence as well as conventional defense to the FRG, making European defense cooperation into a strategic reality. But this decision is as unlikely as another East–West cold spell is undesirable.

A lopsided emphasis in Europe on détente and arms control as against deterrence and defense will be the most vivid expression of that transatlantic crisis for the next ten years. United States calls for a more equitable burden-sharing will invariably (as they have for the past twenty years) fall on deaf ears. Vigorous European participation in the defense of Western interests outside the North Atlantic area will be reluctantly forthcoming at best, especially in the case of West Germany. There are many reasons for this unwillingness to draw the seemingly logical conclusions from the strategic reorientation on the other side of the Atlantic:

- the firm belief among Western Europe's political elites, including to some extent France's, that their countries' security rests mainly on the U.S. presence and the nuclear umbrella associated with it, and that, therefore, an additional increment of national conventional forces does not make much difference;
- those elites' confidence that, in spite of worrisome trends, the United States cannot afford to default *decisively* on her responsibility for the forward defense and nuclear protection of Western Europe for the foreseeable future;
- the fear that rearmament in Europe or involvement in Third World conflicts might endanger détente and crisis stability;
- the insight that a greater Western European role makes sense only in the framework of a European Security Union, which seems out of reach;
- resource-related constraints in countries where the age groups eligible for military service will continue to decline and the transfer of budget funds from nonmilitary to military allocations is well-nigh impossible, except in emergencies, due to the demands of entrenched welfare states;

- the politics of "populism," making elites cater to antimilitary mass preferences in order to carry the next election.

On the other hand, progress in European arms control could still be very difficult. The INF accord exhausts that area of weapon reductions in which both superpowers have a mutual, abiding interest. Deep cuts in short-range and battlefield nuclear weapons, though desirable from the West German viewpoint, will meet with U.S., British, and probably French resistance, and Soviet interest in them will be acute only if they result in a third "zero solution" that effectively denuclearizes West Germany. This has been unacceptable to all in the West except for the Social Democratic Party of Germany (SPD) when it is in opposition. Despite rapid and dramatic progress in the Conventional Forces in Europe (CFE) talks, conventional arms reductions are fraught with enormous complications, to which fourteen years of unsuccessful Mutual and Balanced Force Reductions negotiations bear witness. The consensus in NATO is that European arms reductions should result in "conventional stability," which precludes any but token cuts in Western forces as long as the Warsaw Pact has an offensive capability, because even today NATO hardly meets the requirements for a minimum defense of the Central Front. While the changes occurring in Europe could alter these assessments, the current Western approach to conventional arms control consists of demanding of the Soviet Union that she eliminate unilaterally her invasion potential in Central Europe.[3] Because NATO has little to offer in return, this approach had always appeared futile. However, President Gorbachev's December 1988 announcement of unilateral cuts in Soviet conventional forces, the acceptance of the Pact of many of NATO's demands in the CFE talks (including the need for asymmetric reductions), and the emerging political climate in Europe suggests the possibility of a real movement in this area. For arms reductions beyond those contemplated in the current CFE positions of both sides, it would be necessary for both sides first to realign their force structures so that they become "structurally incapable of attack." NATO could then further reduce its forces symmetrically with the Warsaw Pact, as the Pact could no longer threaten NATO with conventional defeat. Quite apart from the difficulty of finding mutually acceptable definitions of structural incapability of attack, the present force postures must remain in existence for some time simply because present weapon systems may be reduced but will not be written off overnight. Nor will future defensive replacements be rapidly introduced.

Instead of arms reductions beyond those contemplated in the current CFE talks, in the near future the more likely prospect is one of Western European nations not strengthening their defense commitments as the constraints on resources leave them little other choice and the new era of détente seems to render this a permissible course of action.[4] Thus, Western Europe may face an even greater conventional threat for the next ten years even if the Soviets complete unilateral and negotiated cuts. With conventional force improvements such as Follow-On Forces Attack (FOFA) starved of necessary resources, NATO and especially its leading power, the United States, will remain committed to offsetting the Soviet conventional advantage with tactical nuclear weapons, which traditionally have compensated for NATO's conventional weakness.[5] This trend will further alienate West Germany, which, particularly because of the revolution in East Germany, finds itself in a political situation that makes the prospect of fighting a nuclear war in Central Europe more and more untenable.

The crisis of the triad of flexible response, which was never much more than a tenuous rhetorical commitment to quick vertical escalation, is as old as that strategy. Europe, in a slightly self-deceiving manner, fervently clung to that commitment, while the United States has tended, more through her actions than words, to make sure that nuclear escalation could be "controlled." This American tendency has culminated, for the time being, in the removal of INF from Europe and calls for war-fighting strategies along the lines of "discriminate deterrence."[6] Both developments have shocked Europeans, and particularly West German national security elites, who perceive a weakening of deterrence and the apocalyptic vision of a nuclear battle limited to German territory ("singularization"), and insist that the nuclear risk be shared fairly among all NATO partners. The contradiction between the U.S. emphasis on deterrence through *war fighting* and the German emphasis on *war prevention* through deterrence portends even more trouble for the alliance to the extent that there is less readiness in both Europe and the United States to raise the nuclear threshold. For the modernization of theater nuclear weapons in lieu of a conventional buildup (possibly accompanied by deep cuts in U.S. strategic nuclear forces) would only serve to intensify official German misgivings and popular calls for denuclearization. Worse, they could probably be realized only at the cost of another missile controversy and the resurgence in its wake of mass neutralism and pacifism, quite apart from the possibility that they might not be realized at all after the SPD had carried the next elections on an antimissile agenda.

With NATO's conventional strength deteriorating because of tight resources, the modernization of its theater nuclear component in abeyance, and the linkage to strategic levels of deterrence increasingly in doubt, the next ten years do not bode well for NATO's deterrence posture. Radical departures from it, such as the German proponents of "alternative defense" have suggested, will not take place, however. Lacking a feasible alternative, Western Europe will muddle through with flexible response. The cracks in it, though widening like those in the other two pillars of Western European defense, will not lead to its collapse.

EUROPEAN DEFENSE COOPERATION

While European defense cooperation still exists more in words than in deeds, the debate on this issue has intensified over the last years, with much official rhetoric accompanied by some institutional revitalization.

Initiatives in the field of substantive defense policy remain extremely weak, although some significant efforts have recently been launched in arms cooperation. At both political and policy levels the "transnational cooperation approach" is, and will continue to be, the name of the game. This implies the de facto repudiation by all members of the European Community (EC), including France,[7] of the "European approach"[8] and even the "two-pillar approach." Not a European alternative to NATO, or an independent European structure inside the alliance, but closer bilateral and multilateral cooperation between European NATO members is being sought, and has sometimes been achieved, especially in Franco–German consultation and in arms production.

At best, Western Europe attempts to establish a consensus-building framework other than NATO. The most recent endeavors occurred in 1984 and 1986 through two French initiatives to revive the Western European Union (WEU) of the six EC core members plus Great Britain, and through the Genscher–Colombo plan of 1984 to put defense issues on the agendas of the European Council of Ministers and European Political Cooperation (EPC)—the regular sequences of meetings by the EC foreign ministers and heads of state or government, respectively. These two proposals were not put forward in isolation—in fact, the French countered the Genscher–Colombo plan, whose institutional stress on EPC they considered mistaken, with their own WEU solution. Four years later, one may conclude that neither initiative amounts to much. There is no European pillar in the making. Hopes—or apprehensions—about more vigorous European security cooperation at the expense of NATO, in matters such as alliance policies, arms control, strategy, and tactics are clearly belied by the facts.

In particular, the much-touted revival of the WEU has so far proved a failure. Western European Union documents regularly vent their frustration with the near absence of substantive consultation and coordination between member states when there is a real need for them, as most recently in the summer of 1987 on European action in the Persian Gulf crisis.[9] In 1987–88, not only the French but a wide variety of Western European politicians—among them Sir Geoffrey Howe, Hans-Dietrich Genscher, Alfred Dregger, and the Dutch Foreign Minister Hans van den Broek—have called for the drafting within the WEU of a European concept of common security interests, conventional rearmament, and arms control. These calls regularly invoke the "vital importance of the WEU in strengthening the European pillar of the Alliance."[10] And who can question the urgency of collective European action in those fields at a time when the U.S. role in the defense of Europe is becoming less reassuring? In fact, the WEU has so far accomplished nothing in those respects, and there is little prospect that it ever will. For, in one breath with their high-sounding two-pillar rhetoric, European politicians normally state revealing qualifiers such as "the only restriction . . . is that decisions concerning defense policy and the military will be made by the alliance in the future, too."[11] How, under that restriction, can the WEU play any meaningful role at all? The following assessment of the WEU's current and future standing is therefore very much to the point:

> Timid attempts so far, such as the revitalization of the West European Union (WEU), are bound to prove inadequate given the dramatic pressure exerted by the present problems. One cannot revive at will a political mummy like the WEU. . . . Interestingly enough, the efforts to revitalize the WEU have met with the interest of the diplomatic world but not with that of the defense politicians and the military. There is really only one way out of the dilemma: the establishment of a Western European security union which—closely intertwined with the NATO-structures—possesses a military arm of its own.[12]

In view of the European states' inability to get their act together inside a noncommittal forum like the WEU and the perennial lack of progress in nonmilitary fields of European political integration, a security union, which even according to the official timetable would represent the last step toward European unification, must remain an illusion for the indefinite future.

Nor does the alternative of substantive defense cooperation between the major European powers—France, Great Britain, and West Germany—hold much promise at present. Apart from being desirable in itself, this cooperation should in the opinion of its proponents mainly provide for some European supplement to the increasingly unreliable U.S. nuclear umbrella. The German parliamentarian, Jürgen Todenhöfer of the CDU, has proposed an integrated European nuclear force giving the FRG a finger on the nuclear trigger if not her own nuclear weapons (which would remain restricted to France and Great Britain).[13] Franz-Josef Strauss suggested a similar scheme before his death.[14] Ideas like that have a long tradition inside the Gaullist persuasion of the West German right. In 1984 the former Federal Chancellor, Helmut Schmidt, suggested in the *Bundestag* a Franco–German defense cooperation in which a division of labor would extend French nuclear deterrence to West Germany, and the latter would bear the brunt of conventional defense.[15] He has repeated this proposal several times since.[16]

Except for the Social Democratic–Liberal Alliance, which is openly in favor of a fusion of the British and French nuclear deterrents, official as well as unofficial political opinion in those two countries is adamant in its insistence on independent national deterrents. This fact should be borne in mind in assessing the limited technical cooperation that appears to be developing for those two countries' nuclear forces. In March 1987 the defense ministers of Great Britain and France came out of their meeting with the announcement that "we have decided to work together on the evaluation of enemy defences that our nuclear forces have to cross."[17] There are also negotiations in progress about the joint development of an air-launched cruise missile with a range of about 300 miles.[18] After the recent summit talks with France, the United Kingdom suspended her long-standing ban on port calls by French nuclear submarines.[19] More important is the French decision, taken at the same Franco–British summit, to allow reinforcements for the British Army on the Rhine to cross France in case of mobilization.[20]

These are very modest initiatives. The more ambitious proposals regularly run out of steam when it comes to the deliberation of their funding. Nuclear cooperation, whether by pooling France's and Britain's nuclear forces or by extending deterrence to West Germany and the smaller European states, founders on the logic of those two countries' minimum nuclear deterrents. Such a deterrent can become operational only at a time of the most dire threat to national survival, for which purpose it was conceived. This very limit rules out extending it to other nations, or conceding to them a share in its control.[21] Besides, West German treaty obligations forever to renounce nuclear weapons make dual-key arrangements or more elaborate forms of nuclear decision-sharing problematic at best.

In the final analysis it is political discrepancies that stand in the way of genuine security cooperation among the major European powers.

> It would be more than mistaken to talk of a triple alliance London–Bonn–Paris in matters of security. All too frequently coalitions are formed of two countries against the third, or

> in collaboration with the so-called small countries, in order to prevent the supposedly necessary predominance of some individual state. Great Britain still seems to adhere to the continental "balance of power" theory. France appears concerned about the internal and inter-German developments on her eastern perimeter, and the Federal Republic fears, in spite of her economic power, playing second fiddle in security policy and being reduced to the function of "Germans to the front."[22]

Consequently, in matters of security policy, nuclear deterrence, tactics, and arms control, Western European states have preferred either to follow the leadership of the United States and NATO, however much they may grumble at its inconsistency, or they have succumbed to their national instincts, which are actually diverging. Witness, for example, the deep division between West Germany on the one hand and Great Britain and France on the other, over the issue of modernizing short-range nuclear forces and, in general, the desirability of further nuclear arms reductions in Europe; the West German refusal to follow most of her European partners in sending warships to the Persian Gulf; the dissonant reactions by European governments to the U.S. attack on Libya in 1986; and Denmark's recent hardening of her antinuclear policy in defiance of previous understandings with Great Britain and the United States. Only in constantly failing to increase defense spending can the Western European states be said to be truly "cooperating" on defense policy.

Even at the level of theory, at which point independent European efforts have largely ended, the tone has been remarkably atlanticist. A recent EUROGROUP Communiqué of November 30, 1987, stated that "the basic intention of [the WEU] consists in the strengthening of the Alliance as a whole."[23] Of course, the EUROGROUP is a NATO body, and this contention should not, therefore, be surprising. And France, for all her rhetorical support for European security cooperation through the WEU, has, in fact, for years been engaged in a political rapprochement with NATO and, through such efforts as the *Force d'Action Rapide* (FAR) and the French–German brigade, is getting more committed to forward defense—if not flexible response. Referring to France's political reconciliation with NATO, one source has stated: "It is only because of this change in official French security policy that the current debate on reviving the Western European Union can take place."[24] This puts things in perspective very well. With President Francois Mitterand, who is a more reliable friend of NATO and the United States than most of his conservative opponents, in power until 1995, a reversal of France's rapprochement with NATO is not to be expected for the foreseeable future.

France's motive in edging closer to NATO has been not so much to win German consent to the WEU, nor to please the United States, but rather to reassure a West Germany which she sees drifting toward neutralism. This has resulted in a new security dialogue between the two countries at the highest level. Established in February 1982 and building upon the unimplemented strategic cooperation clauses of the 1963 Franco–German Elysée Treaty, this dialogue has culminated for the time being in the establishment at last January's summit of a permanently staffed "Franco–German Defense and Security Council," which is to meet at least twice a year and consists of the two countries' heads of state or government and chiefs of staff of the armed forces (called inspector general in the FRG).[25] Within this framework France and the Federal Republic will jointly draw up an analysis for a comprehensive NATO concept for disarmament, to be completed by 1989, according to an announcement by Foreign Minister Genscher on May 20, 1988.

While France has reassured the Federal Republic politically—a point dramatically highlighted by Mitterand's endorsement in front of the *Bundestag* of the INF deployment—the tangible results of the security dialogue in the military sphere have not been overwhelming. The underequipped FAR's *possible* role in forward defense, the common exercise "Cheeky Sparrow" in November 1987, and the mixed brigade agreed upon at the January 1988 anniversary Elysée summit represent its only fruits so far. Nothing else is as yet scheduled to follow over the coming years. The mixed brigade, its personnel of 4,200 made up of French and German soldiers in equal parts, will from 1989 on be stationed far in the rear of West Germany at Böblingen under the rotating command of a French and German officer.

These efforts are largely symbolic in view of their limited scope, communication problems between officer corps generally ignorant of each other's language, and unanswered questions concerning national prerogatives in the use of mixed units and French participation in forward defense. The latter problem is particularly vexing for West Germany, because its main motive for defense cooperation with France is to "entangle" more and more French forces into the defense of the German homeland against a Soviet attack. The official role of French forces on German soil, including the mixed brigade, is that of "operational reserves." The French government can easily withhold them. This is not entirely satisfying for West Germany. Worse, those forces' capability is subject to doubt. France's emphasis on strategic nuclear forces to the detriment of all of her conventional forces, including those meant for use in West Germany, is undiminished.

Moreover, the role of French tactical or "prestrategic" nuclear forces remains ambiguous, with France only promising to consult the West German government before their employment. In view of the ranges of the *Pluton* and the even more capable *Hades*, their use will largely be limited to either West or East German territory. Both options are unacceptable to West Germany, particularly because there can be no linkage to the use on West Germany's behalf of French strategic nuclear weapons. If the Soviet attack stops on the Rhine, France will not launch them, and there are no signs that this nuclear "sanctuary" strategy of massive retaliation, however hollow it may be, is undergoing significant change.[26] Indeed, it cannot, because that strategy is the only one a minimum deterrent such as France's can implement. In sum, Franco–German defense cooperation can never replace or even decisively supplement the critical role of NATO and the U.S. nuclear umbrella in the defense of West Germany. It should also be mentioned that even the limited cooperation that exists already irritates other European NATO partners, particularly Italy.[27] Nevertheless, the tangible progress in the Franco–German subfield of European defense cooperation does represent a bright spot in the general picture.

The same may be said about only one other field, that of armaments cooperation. The major institutional framework here is supposed to be the Independent European Programme Group (IEPG), consisting of the EUROGROUP plus France, but this group has not yet managed to get a cooperative project underway. While there are IEPG staff goals for five projects (i.e., medium-range surface-to-air missiles—MSAMs; an active, immersible sonar device; light support weapons; coastal minesweepers; and microwave landing systems and ground combat and aircraft electronics), the only step announced so far was the commissioning of a study on the MSAM.[28] Cooperative projects in the armaments field so far have either been channeled through NATO (EUROGROUP and the Conference of National Armaments Directors—CNAD), or they have been subject to bilateral or multilateral cooperation between governments establishing industrial consortia (as in the case of PANAVIA making the three-nation fighter-bomber *Tornado*). But, there has been strictly European cooperation in arms development and production. There are at present two principal projects which hold limited promise that this will become a powerful trend: the European Fighter Aircraft (EFA), or "fighter 90," which on May 16, 1988, the governments of Great Britain, Italy, Spain, and West Germany, against strong U.S. opposition, agreed jointly to develop and build by the late 1990s; and the Franco–German antitank helicopter, agreed upon in November 1987 after years of haggling. The ambitious idea of a Franco–German intelligence satellite in space, promoted by France in order to ease European dependence on the United States for strategic data, has for the moment little chance of realization because of West German misgivings that are political as well as financial in nature.

The other side of the coin is national bickering in cooperative projects, failure to reach cooperation on key projects (such as the Franco–German tank and the NATO frigate, of which fifty units were to have been built in eight NATO countries), a mass of national projects that dwarfs the cooperative sphere even now that EFA is on the drawing boards, and a preoccupation with the two-way street in the transatlantic weapons trade rather than with the intrinsic merits of European arms cooperation such as economies of scale. Nevertheless, arms cooperation currently represents the most powerful and promising aspect of European defense cooperation.

What does the future hold for European defense cooperation? European governments will continue to put their trust in NATO and the United States out of strategic necessity. The only development that could get European defense cooperation moving toward the avowed goal of a security union is a French decision to advance her nuclear as well as conventional defense to the inter-German border, thereby guaranteeing a degree of nuclear protection for West Germany and the smaller Western European states, and for the first time in postwar history relieving the United States of part of the raison d'être for her presence in Europe. This decision remains highly unlikely, especially in light of Eastern European and East German events. And, even if officially declared, the extension of the French deterrent would lack credibility in light of France's insufficient capability to assure its implementation. France and England together might be capable of implementing it, but a fusion of those two countries' nuclear forces for the common defense of Europe is even more unlikely than a French decision unilaterally to extend deterrence. European defense cooperation will thus continue to make agonizingly slow progress well below the threshold of an integrated Western European defense. The three fields in which that progress will occur are Franco–British consultation on nuclear force planning, where it will be most precarious; Franco–German defense cooperation, where it will be most symbolic; and arms production.

ALTERNATIVE OR NONPROVOCATIVE DEFENSE

For the past decade an intense debate about military alternatives to NATO's strategy of flexible response and its operational corollary, forward defense, has rocked the West German defense consensus. Unlike the contemporaneous debate in the United States, whose tone was set by the political right with proposals for offensively oriented strategies against the Warsaw Pact in Central Europe,[29] the debate in Germany centered on initiatives from a nuclearphobic, détente-minded, and frequently neutralist left for a nonnuclear minimum defense. The pertinent military schemes, mostly put forward by "counterexperts" from West Germany's left-leaning "peace research institutes," are commonly summed up under the labels of "alternative defense" or "area defense." Their influence on West German strategy and force planning has hitherto been nil, but, with alternative defense ideas exerting great influence inside the SPD and the changes in the Eastern bloc that may change, especially under an SPD government.

All alternative defense schemes claim to be nonprovocative or, more to the point, structurally incapable of attack. Indeed, besides their nonnuclear character, that is their main claim to fame. It simply means that the alternatively structured armed forces, while allegedly capable of denying an invader the success of his attack, are themselves utterly incapable of attacking at both the strategic and the operational levels (tactically there obviously is no such thing as structural incapability of attack). By thus breaking the "security dilemma,"[30] the concept of structural incapability of attack has important implications for international relations and arms control, and only in this context does it acquire its relevance. According to one of its main proponents, Dieter S. Lutz,

> structural incapability of attack is not an independent concept or strategy either in respect of its political implications or from the perspective of its chances of realization. It must rather be seen as an integral part of the larger political concept of "common security.". . . In the short and medium terms, common security aims at
>
> - the realization of structural incapability of attack
> - the replacement of deterrence
> - arms limitations and disarmament through negotiations and also through unilateral measures.[31]

In what follows, the multifaceted and problematic set of ideas of the alternative defense school will first be covered, and then the concept of structural incapability of attack will be dealt with, emphasizing its consequences for deterrence and arms control.

Alternative Defense

Most simply put, alternative defense is area defense by infantry as opposed to forward defense by armor, without the nuclear option of the latter. It is essentially *conventional deterrence by denial*—"conventional" because it dispenses with battlefield and theater nuclear weapons and "by denial" because it deters by promising to stop the attack rather than ultimately punishing the attacker (as in flexible response). Being, moreover, structurally incapable of attack, alternative defense not only deters the malicious aggressor but also restrains the would-be preemptive aggressor. The secret of its widespread appeal lies in this combination of nonnuclear defense with an allegedly better-than-nuclear deterrence.

It is indeed no accident that alternative defense ideas have been thriving at a time when, because of the heated INF controversy, an acute awareness of the nuclear dilemma inherent in West Germany's defense has reappeared among the larger population. There is no better argument for alternative defense than the existing strategy itself, entailing as it seemingly does the nuclear destruction of the very things it is supposed to protect. Throughout postwar history, alternative defense thinkers have enjoyed influence and appeal whenever the nuclear dilemma was on the people's minds and have been in limbo whenever defense issues were of little concern to the populace. Thus, it was in the mid-1950s, when the current defense strategy was taking shape and dividing the West Germans over the issue of a nuclear-capable *Bundeswehr*, that the "grandfather" of all alternative defense proponents, Colonel Bogislaw von Bonin, gained prominence. He suggested for the new German Army a two-tiered force structure of light infantry and armored reserves bereft of nuclear options, and he was immediately dismissed from his planning post in the defense ministry.[32] The tradition he founded was resumed in the early- to mid-1970s by Carl Friedrich von Weizsäcker and Horst Afheldt at the Max Planck research institute at Starnberg. In 1976 Afheldt published his *Verteidigung und Frieden*,[33] which has remained the classic among the vast alternative defense literature. But throughout the seventies, both the debate about, and the literary contributions to, alternative defense were in much greater evidence outside the FRG than inside that country. A veritable debate about the merits of territorial defense and light infantry forces for NATO took place in the United States.[34] Austria officially introduced an area defense, for which General Emil Spannocchi had laid the theoretical groundwork.[35] In France, General Guy Brossollet created quite a stir with his proposal to reduce French defenses to massive nuclear retaliation plus a lightly armed area defense.[36]

In the FRG, alternative defense concepts would have remained the esoteric business of a few defense intellectuals if political deviators from the West German defense consensus had not discovered them "punctually" in 1980, years after their original formulation. With the INF decision of December 1979, the clouds of nuclear annihilation appeared to many Germans to be drawing close. This perception was quite apart from the potential alternative defense proffered for the promotion of détente, neutralism, and antidefense sentiment pure and simple, all of which were then bursting forth in unprecedented strength. It is ironic, therefore, that the West German debate was to be kicked off by someone who had nothing to do with those particular motives. General Franz Uhle-Wettler, an upright German *Panzergeneral,* in early 1980 published his *Gefechtsfeld Mitteleuropa: Gefahr der Übertechnisierung von Streitkräften* (*Combat Zone Central Europe: The Danger of Oversophistication of Armed Forces*),[37] in which he recommended changing the army's force structure to light infantry *where the terrain strongly suggested it* (i.e., in the growing urban and wooded sections of the inter-German frontier). There, he quite rightly said, armored forces would be wasted. Several articles in *Der Spiegel* gave that technical argument wide publicity and (mis)used it to sow general doubts about the structure and strategic orientation of the *Bundeswehr*.[38] Light infantry became fashionable, and, in a new politico–strategic climate, so did the almost forgotten ideas of Afheldt, Weizsäcker, and others. They could come into their own under the post-1980 maxim of security policy on the West German left, that "it is not the enemy but the [nuclear] hostility itself that is the main danger."[39]

A description of the historical background of alternative defense would be incomplete without mentioning previous military experiences that seemed strongly to support this idea. The alternative defense paradigm represented a theoretical novelty only in the context of the Central Front and of Austria. Home-grown versions of area defense by infantry forces had long ago been adopted, and successfully tested, by neutrals such as Finland (the "Sissi-Motti" tactics that inflicted heavy losses on the Soviet invaders in 1939–1940), Yugoslavia (Tito's partisan war), and the People's Republic of China (Mao Tse-tung's people's war) and its epigones in the Third World, like Vietnam. Moreover, the experiences of the U.S. Air Force in Vietnam, the Israeli tank forces in the Yom Kippur War, and the Soviet army in Afghanistan seemed to suggest the limited utility of heavily equipped troops against large numbers of foot soldiers with modern antitank and antiaircraft weapons.

Quite logically, therefore, the latter-day theoreticians of area defense have combined the guerrilla tactics of those previous examples with high-tech antitank weaponry, especially precision-guided munitions (PGMs) and other emerging technologies if they satisfy the requirement of structural incapability of attack. This combination forms the tactical essence of alternative defense. Strategically, it is to achieve

- conventional deterrence by denial;
- successful defense against a mechanized attack if deterrence fails;
- protection of the object of defense, or damage limitation;
- crisis stability by removing incentives for preemptive attack;
- "social compatibility," that is, low costs and a force makeup in accordance with democratic norms.

In what follows, the central features of alternative defense that are common to all individual proposals shall be laid out in reference to those five goals.

First, alternative defense is conventional because it entirely dispenses with battlefield and theater nuclear weapons on West German territory. It relies implicitly or explicitly on sea-based strategic nuclear forces to provide the minimum deterrence of nuclear blackmail still considered necessary, or eliminates nuclear deterrence from the equation altogether (only a minority—leftist experts close to or in the Greens—advocate the latter stance). It is necessary to point out here that everybody in West Germany, from right to left, rejects the idea of fighting a nuclear war in Central Europe, even as a deterrence posture. Even conservative opinion in the FRG concedes to nuclear weapons strictly the political role of preventing war, and refuses to rule out a Western first use within the theater only because otherwise that political deterrent effect would be undermined. The U.S. nuclear forces on German soil, according to that opinion, exist mainly for the sake of "coupling," at most for releasing a nuclear warning shot or two preferably onto Soviet territory, but by no means for war fighting. Alternative defense, apart from finding the very idea of nuclear deterrence repugnant, essentially wants to make totally sure that this opinion can prevail in the reality of war. It wants to denuclearize West Germany and deprive her defense of all targets attractive for nuclear strikes in order to minimize the possibilities of nuclear escalation by either side. Alternative defense thinkers tend to see the danger of nuclear escalation emanating more from the very presence of nuclear weapons and delivery systems—which the enemy will seek preemptively to destroy and friendly forces will hurry to use rather than lose in the first crisis of battle—than from the breakdown of friendly conventional defenses.

To be sure, those thinkers do not believe that forward defense will long remain conventional. Today's deterrence is said to be essentially nuclear—by punishment. They consider it a mistake to confront the attacker on his own terms (i.e., with tank forces which his superior tank armies are geared to penetrate and defeat). This permits Soviet hopes for a quick and decisive victory, reducing crisis stability in peacetime and conjuring up immediate nuclear escalation in wartime. Denuclearization must thus be complemented by a promising scheme of conventional deterrence by denial. A conventional arms buildup is ruled out because it would lead to political tensions, to unacceptable social and financial costs, and, by sharpening the security dilemma, to less crisis stability in Central Europe. Moreover, conventional forces of the present configuration, which present a mass of attractive targets for nuclear strikes, could, if they were to prove steadfast, by their very success provoke the nuclear escalation which alternative defense proponents so dread.

The answer lies in the *reductio ad absurdum* of deterring both conventional and nuclear attack through area defense.

Area defense consists of a theater-wide network—or, in the more recent proposals, a forward defensive belt—of light infantry, artillery, engineers, obstacles, air defense units, and in recent proposals, a few armored reserves. (Major individual proposals will be covered in more detail below.) All have in common a reliance on dispersed "shield" forces of light infantry which, with or without the backing of armored "sword" elements, are to engage the tank-heavy attacker in an area-wide battle of attrition that should result in his repulsion or decimation before he can reach his objectives. Space may be given up if the offense will thus be reduced *ad absurdum*. Because the would-be attacker will foresee this outcome and because the use of nuclear devices to turn it in his favor makes no sense against the dispersed and concealed troops in the defensive networks, launching an attack is futile. The idea is to convince the potential aggressor that he can attain his objective only at a prohibitively high cost, if at all; or, conversely, that the objective can never be worth the "entry fee" and "residence fee." Alternative defense deters primarily by denial but it retains the cost–benefit calculations of deterrence by punishment, albeit without (nuclear) costs for the enemy's population and territory: a high cost is inflicted on the invader, but alternative defense does not punish the invader by attacking his industrial base or population.

As to the second goal, why will that battle of attrition be successful? Alternative defense proponents argue that in our days of rapid technological progress, defensive developments have called into question the superiority of the offense—"fire defeats movement."[40] High-technology weapons, particularly precision-guided munitions, are said to give the dispersed defender a decisive advantage over the armored attacker, above all in the increasingly compartmentalized terrain that has been created by urbanization and afforestation in West Germany. Elusive light infantry units will not allow themselves to be drawn into pitched battles, where the enemy can bring to bear his superior firepower. Only rarely will they hold the line—in contrast to traditional notions of defense by infantry. Having surprised the attack formation with massive fire from their initial positions, they will rather seek to delay a superior enemy and attack inferior opposition inside preassigned combat zones. They will always avoid presenting targets for massive strikes, but they themselves will constantly strike at the enemy's massive targets (tank formations, assembly areas, march and supply columns, and the like). In the end they will have taken a toll that forbids the enemy to continue his offense. Many proposals then envisage the use of armored reserves to deliver the death blow and expel the remaining hostile troops from the territory of the FRG. Should that not succeed, the area defense will at least have held out long enough to prevent a fait accompli, buying ample time for diplomacy to restore the status quo ante and—again most important—rendering nuclear escalation less urgent or altogether unnecessary.

Third, if deterrence fails, area defense will also succeed in damage limitation. No nuclear devastation can occur. For, irrespective of the success or failure of the area defense, even irrespective of whether or not nuclear escalation occurs, a West Germany deprived of targets for nuclear strikes will be spared. The Germans, as it were, might in the worst case see the missiles fly back and forth over their heads but will never be hit by them. Collateral damage in area-wide conventional warfare will be small in the absence of "decisive" battles. The thousand skirmishes between the infantry parties and hostile units will be of short duration and comparatively low tactical intensity, their cumulative effect being destructive to the enemy but not to the country. The clear-cut combatant status of the light infantry will protect the civilian population from subjection to reprisals and other methods normally considered appropriate in countering a partisan-like enemy. In sum, the combat would limit damage to West German people and property, avoiding the element of self-deterrence so prominent in an official strategy predicated upon nuclear escalation.

Fourth, crisis stability has been defined as follows:

> Neither side must be able significantly to improve its position by striking first in a crisis. There must be a pay-off for waiting, and there must be no pressure exerted by the military instruments themselves for their quick or preemptive use. The structure of the armed forces must be such that mobilization races do not occur during the crisis. Beyond that there should be no attractive targets with the defender which the enemy can quickly and possibly preemptively destroy.[41]

Alternative defense tries to meet those requirements quite thoroughly. It claims to be structurally incapable of attack, thereby assuring the enemy that preemption is unnecessary. It eliminates all weapons that are attractive targets—by reason of their importance and simultaneous vulnerability—for preemptive first strikes, including land-based nuclear forces, heavy attack aircraft on the ground, large armored formations, and all concentrations of forces. Nor does it retain any weapons which come best into their own through preemptive use. To a larger or lesser extent, it is dependent on mobilization, but the decision on mobilization may be taken early without aggravating the crisis because the force posture remains nonthreatening even after full mobilization.

Finally, alternative defense is cheap, or so its proponents say. It undoubtedly does away with those weapon systems that traditionally have cost the most money (e.g., tanks, heavy artillery, and attack aircraft). No alternative defense proposal raises significantly the personnel strength of the *Bundeswehr*—in fact, many stay well below it. The only thing that increases is the reliance on reservists, the costs of whose sporadic unavailability to the civilian economy are counterbalanced by the shorter active duty service which most proposals envision. Most schemes also cut down on the number of volunteers and professional soldiers—on this Afheldt is idiosyncratic as always in suggesting an area defense of volunteers only. Far from militarizing society, this inexpensive militia army will democratize the military. This is a favorite theme of the decidedly leftist protagonists of alternative defense.

All alternative defense schemes are virtually homogeneous in how they intend to achieve the goals of conventional deterrence by denial, damage limitation, crisis stability, and social compatibility. They differ widely, however, on the configuration and, especially, the tactical emphases of the area defense. In this respect, alternative defense proposals may be classified by the degree to which the area defense remains concentrated forward, retaining elements of forward defense, in conjunction with the degree to which it retains mobile armor and thus today's option of operational counterattack. Four categories emerge:

- No concentration forward—no mobile armor: this is the original and most popular articulation of alternative defense. It represents "pure" area defense by light infantry. Its author is Horst Afheldt; Günter Vollmer has further developed it.[42]
- All forces concentrated forward—almost no mobile armor: this is Norbert Hannig's "fire barrier" concept; it has been further developed by Johannes Gerber.[43]
- Area-wide defense concentrated forward—weak mobile armor: this is the model of the Study Group on Alternative Security (SAS).[44]

- Area-wide defense concentrated forward—strong mobile armor: though far less popular than pure area defense à la Afheldt, most authors of alternative defense schemes actually have chosen this combination. Most important among them are the SPD defense expert, Andreas von Bülow, and Jochen Löser, Albrecht C. von Müller, and Lothar Penz.[45]

Let us consider each in turn.

First, Afheldt's area defense has become known by its key element, the "techno-commandos." In his 1978 classic these light infantry commandos, supported by missile artillery, were all the area defense consisted of; in his 1985 work, he concentrates them in a forward belt between 50 and 100 kilometers in depth, with mechanized forces and missile artillery behind them, but this is to him only a "transition solution" pending the consummation of pure area defense.

The techno-commandos are light infantry platoons of about twenty-five men, adding up to a field army of about 470,000 (including combat support and relatively weak supply troops) which covers the whole territory of the FRG. There is no air force. In combat, each techno-commando is constricted to a preassigned combat zone of 10 to 15 square kilometers. Here, the light infantry platoon fights enemy armor that happens to be passing through with short- to medium-range PGMs, which the soldiers fire indirectly from field fortifications, the PGMs being self-seeking devices of the third generation. Thereafter, the platoon engages soft enemy targets with less fanciful weapons in classical "hit-and-run" light infantry style. Shoulder-held antiaircraft missiles provide protection from air attack. Longer-range missile artillery (with ranges up to 80 kilometers) and some concentration of techno-commandos along the likely axes of attack create main defensive efforts at the operational level. This makes up for the complete lack of both operational and tactical mobility in Afheldt's light infantry army, which makes it truly into a force structurally incapable of attack. Nor is there any command, control, and communication (C^3) to speak of. Each techno-commando platoon fights on its own, with C^3 limited to the employment of the heavy missile artillery and communication of reconnaissance about enemy concentrations to, and between, techno-commandos on the spot. Between stationary techno-commandos, communication can proceed by field telephone connections, which are much more resistant to jamming and signal reconnaissance than the radio communication necessary for mobile warfare. Afheldt's is an army of self-sufficient high-tech guerrillas.

Vollmer, improving on Afheldt's ideas, finds the number of troops in the model too small for the task. "The number of defenders is set too low in the model. Invading mass armies will not be held up with roughly 400,000 German and an undetermined number of Allied troops."[46] He also introduces more tactical variety into Afheldt's mono-tactical method of indirect fire with third-generation PGMs.

Second, Hannig's fire barrier consists of a massive belt along the border of short-, medium-, and long-range PGMs, with ranges from 6 to 200 kilometers, and mines. This barrier is supposedly capable of preventing an armored breakthrough at any given point in the front. The attacker is saturated with PGM fire from overlapping, multiple-range launching positions which are deployed in echelons according to their ranges. Electronic sensors and terminally guided third-generation missiles form the high-tech backbone of the system. The launchers gain survivability and staying power from entrenchment, field fortifications, and armored protection. The fire can extend well into the enemy's rear area beyond the border, but the complete lack of mobility in Hannig's concept leaves the fire barrier structurally incapable of attack. Ultrarapid concentration of fire by PGMs of all ranges substitutes for mobility. Gerber improves upon the concept by adding weak mobile elements (armor and antitank helicopters) for the liquidation of forces that unexpectedly manage to break through the fire barrier. Figure 1 shows how the Hannig–Gerber model is supposed to work.

Third, the SAS model of an "interactive forward defense" moves Afheldt's territorial area defense into a forward belt and adds to the stationary light infantry commandos some mobile armored support as well as (antitank) helicopters. The ratio of mobile to stationary units is 1:3.9 on the basis of troop strength. The infantry fights according to light infantry tactics in combat zones of approximately 8 square kilometers per twenty-eight-man platoon (60 square kilometers in a screening belt in front of the main defensive belt) and relies much less than Afheldt's techno-commandos on fanciful high-tech weapons, whose costs and complexity the SAS finds prohibitive. A robust mix of mines, obstacles, mortars, automatic canons, and short-range antitank missiles forms its armament. Light tank units on the lines of today's armored cavalry would back up the infantry in close terrain, while heavy armored forces would do so in open terrain, being also ready to regain lost territory through counterattack. "Close interaction between static and mobile elements . . . could prove advantageous. . . . The area-covering system would wear down the aggressor's strength, canalize his movements, and serve as a source of intelligence as well as cover for our own counteroffensives."[47] A thinly spread network of light infantry provides rear-area security in the SAS scheme.

Fourth, Bülow ("army structure of the nineties"), Löser ("area-covering defense"), Müller ("integrated forward defense"), and Penz have taken up the old theme of the two-tiered force structure which Bonin introduced into the discussion more than thirty years ago. Subsequently, defense thinkers such as John C. Slessor, B. H. Liddell Hart, and Ferdinand O. Miksche proposed much stronger two-tiered deployments for NATO's Central Region, arguing that only large militia forces could supply the numbers of troops necessary for an effective defense and release the armored forces for their natural task of counterattack.[48] Bülow, Löser, and others follow those three thinkers, adding the alternative defense rationales, in their advocacy of a rather powerful two-tiered structure for the West German army. Their proposals by and large maintain what was the officially planned peacetime strength of the *Bundeswehr* for 1995 of about 450,000 men, but considerably increase the number of combat brigades after full mobilization (within forty-eight to seventy-two hours), although that increase is in infantry only, and all but some of today's mechanized brigades would be converted into light infantry brigades. Löser, for example, doubles the number of combat brigades after mobilization.

Löser's defense scheme is well thought through. This retired major-general suggests a forward defensive belt 40–60 kilometers deep, where the West German

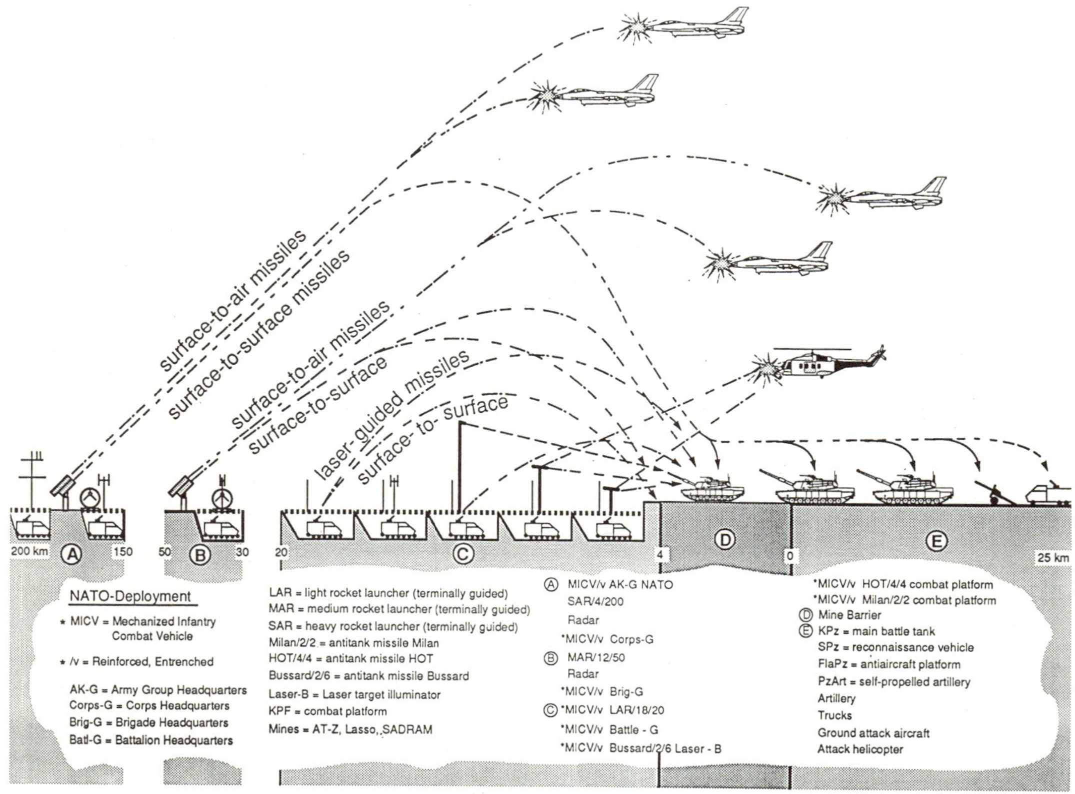

Figure 1. The defensive action in the forward defensive belt takes place from reinforced field positions and full cover. This proffers the defender the better protection than does armor for the attacker (redrawn from *Sicherheit und Frieden* 1, 1987, p. 30, by permission).

light infantry brigades—the bulk of the army—conduct fluid antitank warfare in accordance with light infantry tactics. The *terminus technicus* in German is *Jagdkampf*. Hostile armored forces will be channeled by the terrain and blocking devices and forced to slow down and amass before numerous artificial as well as natural obstacles. Being thus caught in "fields of fire," they will be subjected to concentrated antitank fire from the troops overwatching the obstacles as well as from several ambushes in their flanks and rear. In this the light infantry makes use of a robust mix of PGMs and conventional weaponry. Longer-range PGMs will engage the bogged down enemy with indirect fire—"Concentration of fire smashes the concentration of armor." Light infantry forces temporarily withdraw after the engagement to lure the enemy into further obstacles in the rear and repeat the described combat phase. As in previous proposals, units fight in preassigned combat zones, but the density of soldiers per square kilometer is much greater than in the Afheldt and SAS models, and a hierarchical command and control structure is maintained during combat.

Armored forces stationed behind the forward defensive belt in the so-called area defense belt, where less densely deployed light infantry provides rear-area security, would counterattack against enemy forces threatening to break through and proceed to regain occupied territory from the weakened aggressor. NATO armored forces incorporate all allied units as well as the *Bundeswehr's* remaining mechanized brigades. The ratio of light infantry to armor is 1:1 overall. A light infantry brigade would have 2,800 men organized into three light infantry groups (batallions), plus signal, helicopter, and supply and medical elements. A light infantry group is composed of three light infantry companies, one assault company, one missile company (armed with PGMs), and headquarters and supply elements. The light infantry company consists of three platoons and one ranger platoon. The platoons, in turn, have one infantry squad, one antitank squad, and one missile squad. Figures 2 and 3 illustrate the combat methods of those units.

Bülow's, Müller's, and Penz's concepts are tactically similar to Löser's but display less military expertise in detail. Bülow, as a politician concerned about forward defense, stresses conventional infantry tactics in the forward defensive zone. The unarmored infantry will often have to hold the ground from entrenched positions echeloned in depth. An analogy is the battle of Kursk in 1943, where a deep network of infantry-manned antitank positions succeeded in halting the German armored thrusts. Müller emphasizes high-technology weaponry for the light infantry. He and Penz are inclined to grant the armored counterattack the central role in the defense. Penz's ratio of armor to infantry is close to Löser's, while Bülow's and Müller's is approximately 1:2.5.

Can alternative defense live up to its own claims? First of all, can it achieve conventional deterrence by denial? As will be pointed out presently, alternative defense suffers manifestly from fatal flaws at the operational, tactical, and logistics levels, where its predominant ingredient of area defense comes into play. Does this fact incapacitate alternative defense at the strategic level, too? One could argue, as many backers of alternative defense do, that forward defense is operationally and

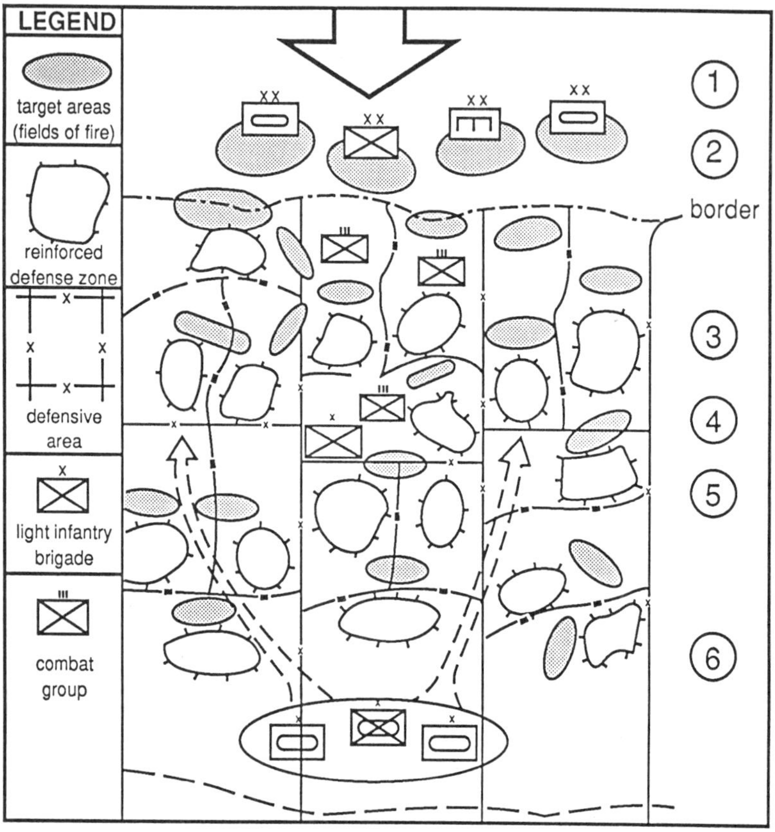

Figure 2. Network of the forward defense of the border area (redrawn from *Sicherheit und Frieden* 1, 1987, p. 57, by permission).

tactically not very promising either,[49] and that alternative defense should, therefore, be preferred over flexible response as the less risky and less costly strategy.

However, the existing operational method of forward defense enjoys the strategic backup of nuclear and geographical escalation, in conjunction with which it enhances deterrence by making a quick and decisive victory *necessary and simultaneously uncertain* for the Soviet Union. For that kind of victory is an absolute must, given the decidedly unattractive alternative of trans-European and possibly nuclear war which flexible response prescribes and thc physical U.S. presence in Europe reasonably well ensures. But nobody can guarantee the Soviets that forward defense, for all its weaknesses, will quickly and decisively collapse. And even if this fait accompli should happen, escalation is still possible.

Alternative defense, on the other hand, explicitly deters only at the conventional theater level (i.e., the operational level), and it is exactly there that its capability is insufficient. It does not envisage any nuclear coupling mechanisms to U.S. strategic nuclear forces, and conventionally, while stopping short of demanding the withdrawal of all U.S. troops from West Germany, its proponents have never revealed how they want to reconcile a continued U.S. presence with demands for denuclearization and radical restructuring which no American administration will ever accept. Most alternative defense proposals explicitly charge only West German forces with manning the defensive networks, implying the ultimate relief of allied troops of their forward responsibilities if not of maintaining a rear or symbolic presence. Nevertheless, deterrence would crucially suffer from the abandonment of the current shoulder-to-shoulder deployment in national corps sectors. This deployment makes it extremely difficult to limit the attack to anything less than a confrontation with the Western Alliance as a whole. Last, but not least, alternative defense undermines deterrence by deliberately depriving itself of the ability to threaten the aggressor's territory with nuclear strikes, air attacks, and conventional counterattacks (only the Müller proposal envisages at least the latter). Thus, as an invitation for a conventional rout of German forces by Soviet armies, alternative defense depends on the goodwill of the U.S. president to bail out West Germany, either through arduous reconquest or through nuclear threats. This is indeed a very poor form of deterrence.

Even if area defense should operationally work the way its adherents claim it would, it still could not sufficiently provide for the reconquest of lost territory or for

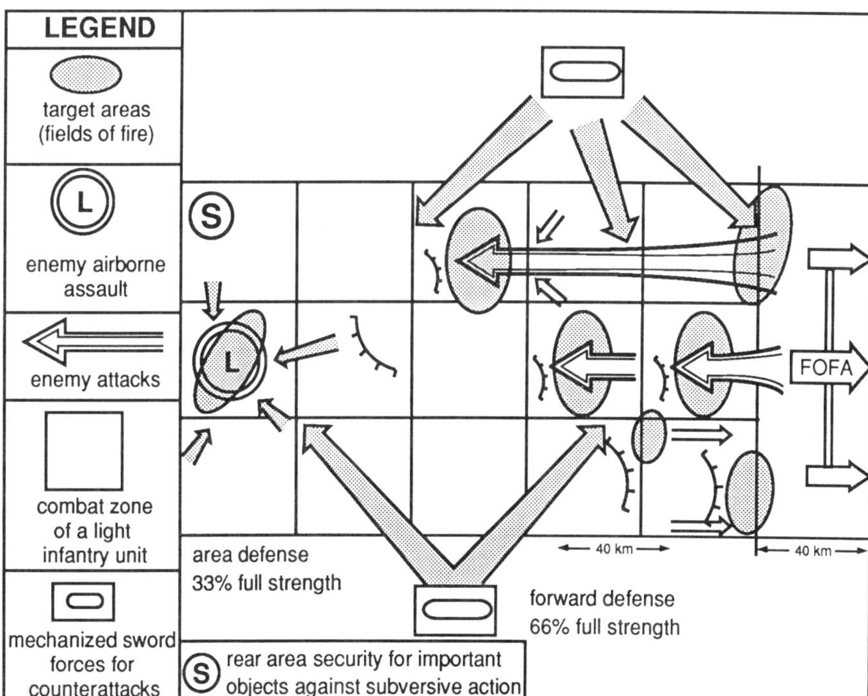

Figure 3. Combat method in the network forward defense of the border area (redrawn from *Sicherheit und Frieden* 1, 1987, p. 57, by permission).

a defeat of enemy forces. It lacks the armored punch and operational mobility necessary for those options (the only exception may be the Löser proposal with its ratio of armor to infantry of 1:1). Despite high expected losses, this fact makes the operational risk calculable for an attacker whose strategic risk is minimal already, and who enjoys a sanctuary in which he can move fresh troops to the front at will. If he is bent on attacking, the light infantry defense cannot deter him from doing so. It would have to take recourse to nuclear weapons, the very thing it was designed to make superfluous.

By minimizing the strategic and operational risks for the would-be aggressor, alternative defense critically weakens deterrence. Conventional deterrence by denial, as understood by the advocates of an alternative and nonprovocative defense, is simply not enough deterrence. Worse, this incomplete deterrence itself caves in on account of its fatal flaws at the operational, tactical, and logistical levels. What are these flaws?

A preponderance of light infantry and static high-tech firepower in the force structure will prove fatal, because the results in battle will be as follows.

Electronic countermeasures, deception and suppression by artillery, and PGM fire by the attacker can disable the defender's PGMs, their launchers, and their sensors, particularly if these are not mobile and are unarmored. An area can be cleared of light infantry by hostile light infantry of superior strength using like combat methods. Or, if small enough, the area can be saturated with massive artillery barrages so that dispersed as well as entrenched infantry will be suppressed. The area in question will be small. For the attacker needs to employ those predictable countermeasures only in his chosen corridors of advance, which in Soviet doctrine are quite narrow at the point of breakthrough (about 10 kilometers). For this task he has an abundance of artillery, and he will always be able to muster the infantry for limited clearing operations. There are also quite unconventional techniques for clearing corridors which will be very effective, such as razing wooded areas to the ground by fire so as to deprive the defenders of any cover.

Operational maneuver groups and tank armies can begin rolling through the corridors long before they are wholly cleared, because defenders whom clearing parties and artillery barrages keep busy will not be able to put up an organized resistance. Hence, penetration will be almost immediate. The offensive thrusts will recklessly advance, duly securing their flanks against light infantry attacks and smashing the defender's weak armor racing to the points of penetration in meeting engagements. The hostile forces will proceed to encircle the vast networks of infantry in the forward defensive belt. If the defensive network extends countrywide (as with Afheldt), the attacker will continue clearing corridors for his operations at will, the task made even easier by the low tactical density of defenses thus deployed. Air attacks will knock out long-range missiles conceivably capable of concentrating fire for operational purposes. *Perhaps 95 percent of the area defense will suffer defeat without ever having seen combat.* Their lack of operational mobility and inability to attack enemy forces renders them irrelevant to the mechanized invader. He can even concentrate *all* his forces at the axes of attack, since there is no need to tie down troops that can neither move nor attack; a thin screen of infantry along the border will be sufficient to guard against them. Nobody can doubt that these massive armored thrusts will cut through vastly outnumbered localized infantry defenses like knives through butter, even without employing special countermeasures. And the absence of air and nuclear threats allows massive offensive concentrations.

Thus, within a few days at best the area defense will have to capitulate or switch to guerrilla warfare behind enemy lines. Quite apart from the logistical questions this raises in an easily controlled environment like the FRG, the idea of Germans fighting a guerrilla war against Soviet occupation is simply preposterous (except, perhaps, in areas like the Hamburg–Hafenstrasse).

The above assessment is less applicable to the extent that a given proposal includes more armor. Against Löser's defense of 50 percent infantry–50 percent armor, for example, the Soviets could obviously not afford to amass all their forces at the axes of attack. They would also face stiffer opposition from friendly armored reserves following the breakthroughs. The option some proposals retain of long-range fire onto enemy territory would serve to inhibit the enemy's operational movements and concentrations. But to a larger or lesser degree, all area defense schemes tend toward the disastrous outcome described above.

Obviously, the main reason for this outcome is the missing ability to create defensive concentrations at the operational level except by easily neutralized long-range PGMs and vastly outnumbered armored reserves. A Warsaw Pact invasion can only be stayed if hostile forces have to grapple with troops that confront them in strength, during the defense and the counterattack alike, wherever and whenever the main offensive efforts occur. Only armored troops are mobile and steadfast enough to meet that requirement. And only today's combined arms concept does not run the deadly danger of technical and tactical unidimensionality to which area defense, with its easily neutralized key elements of PGMs and light infantry, falls victim.

Area defense achieves damage limitation only by its predictably rapid collapse, which is not quite the way in which its advocates want to see it achieved. Should the area defense prove its mettle, however, the opposite of damage limitation would result. It follows from the aforementioned Soviet countermeasures that the area-wide combat, far from lowering the intensity, would only serve to subject most of West German territory to Soviet artillery barrages and airpower. This would shatter the myth of light infantry's immunity to massive firepower and also destroy much of West Germany's vulnerable socioeconomic fabric. Many of the combat methods of the area defense resemble guerrilla tactics. Despite all claims to the contrary, they may well provoke pressure and reprisals against the civilian population.[50] At best, area defense exchanges one form of self-deterrence for another.

If alternative defense were a viable form of deterrence, the removal of incentives for preemptive attack by either side would indeed bolster crisis stability. However, area defense schemes that retain sizeable armored forces present a new problem. Relieved of forward defense duties, these forces are in their "sword" role and, therefore, ready to attack at any time. The force posture becomes structurally even more capable of attack since today most armored units are tied down in forward defense, unable to form offensive main efforts.[51] These conflicting arguments are rather academic because a deterrence that does not work is the worst thing that can happen to crisis stability. For regardless of security dilemmas and 1914 analogies, the first principle of crisis stability is still the old *si vis pacem, para bellum*.

The last question to be answered is whether alternative defense would be as inexpensive and democratic as its proponents claim. Nobody can say in advance whether armed forces that are organized on a local militia basis, highly dependent on civilian cooperation and regularly inducting huge numbers of reservists, would democratize the military or, in fact, militarize the society. Depending on the political climate, it may turn out either way, but the potential for the latter outcome certainly exists. As far as the price of alternative defense is concerned, it would be low only if today's oversophistication were not superseded by another kind of high-technology monstrosity in the form of an army built around PGMs of all sorts. Hannig's fire barrier, for example, would clearly be prohibitively expensive. Nor is it clear whether the direct as well as indirect costs of frequent and massive call-ups could be entirely neutralized by a somewhat shortened active-duty service and reductions in the number of professional soldiers. Both of these measures could not go very far since the peacetime strength of the armed forces does not substantially decrease in the majority of proposals. In sum, an alternative defense based on the large-scale use of PGMs and reservists may well turn out to be more expensive than the *Bundeswehr* today (where, it must be admitted, reliance on reservists will also increase in the future).

In view of the described shortcomings, the future of alternative defense will be in the hallways of peace research institutes and not in the batallions of the *Bundeswehr*. Even in the SPD there is no majority for such a radical departure, although its most prominent defense experts, Andreas von Bülow and Egon Bahr, flirt with a radical restructuring of the *Bundeswehr* and take an active part in the debate. But once in power they and their followers in the party would have to bow to better knowledge. Unilateral denuclearization and adoption of area defense would represent an irresponsibility which, for the foreseeable future no German politician could commit.

Nonprovocative Defense

Structural incapability of attack has been defined as follows: "[It] obtains, in a wider sense, when the social and political organisation of security by potential enemies manifestly and permanently precludes wars as a means of policy. Structural incapability of attack in a narrower sense exists when the armed forces serve the function of military denial and efficient defense but manifestly do not permit military aggression given their organisation, structure, armament and strategy."[52] The concept first cropped up at the Max Planck research institute in 1983 (its staff member Albrecht C. von Müller claims the copyright), but in fact the idea was developed by Carl Friedrich von Weizsäcker and Horst Afheldt at the same institute in the early 1970s and found expression in their aforementioned works. Structural incapability of attack represents the indispensable precondition for a system of common security which aims to replace deterrence with collective security as the preferable means to maintain military and political stability in Europe. In Europe it is the SPD and, rhetorically at least, the Warsaw Pact which subscribe to the goal of common security. Thus, structural incapability of attack, apart from being a trademark of alternative defense, is an integral part of the defense policy of one major West German party. In order not to fall behind the latest fashion in the ongoing West German defense debate, the conservative government has lately included the concept in its vocabulary as well.

So far, however, there is not a trace of agreement as to when a state of structural incapability of attack actually exists. The above definition is obviously a tautology. To fill it with meaning, one would have to specify what *kind* of organization, structure, armament, and strategy met the requirements of the concept. The West German government deduces from the present inability of NATO forces to launch a war of aggression that structural incapability of attack would obtain in Central Europe once the Warsaw Pact had come down to NATO's present force levels.[53] This begs the question whether the Warsaw Pact would then still be satisfied that NATO cannot present an offensive danger (and vice versa, of course). The SPD and allied defense reformers, on the other hand, think that the Warsaw Pact is already uncertain as to NATO's strictly defensive capability. They demand that the West as well as the East undertake steps toward a nonprovocative defense. The Warsaw Pact leaves it at the admission of certain "asymmetries" in both sides' capabilities. Only the proponents of alternative defense pretend to know what the concept means—a restructuring of the armed forces into an area defense bereft of nuclear weapons, heavy firepower, and large mobile strike forces.

It is clear that for the concept of structural incapability of attack to have any meaning, it must refer to a force posture that dispenses with large numbers of weapons capable of striking or invading enemy territory and replaces them with weapons inherently incapable of that. It is in this narrower sense that the concept is generally understood. It is also in this sense that the concept must remain irrelevant for the foreseeable future. For apart from the difficulty of finding mutually acceptable definitions of structural incapability of attack, the present force structures in East and West, even in the event of a CFE agreement, must essentially remain unaltered for at least another fifteen years, simply because present weapons systems and acquisition programs will not be written off, nor future defensive replacements be introduced, earlier than that. Thereafter it may take another fifteen years to consummate the defensive restructuring. Moreover, it is by no means clear whether the Soviet Union can afford a structurally defensive force posture even if she permanently forswears her ability to invade the West. For despite the tolerance of the dramatic changes that have occurred thus far in Eastern Europe, she may indefinitely need an ability to intervene in her Warsaw Pact allies' territories. Finally, structural incapability of attack must, strictly speaking, imply a Europe disarmed of nuclear weapons "from the Atlantic to the Urals" because nuclear deterrence is incompatible with, and to be superseded by, that concept. However, despite the processes of change that have been unleashed in Eastern Europe, such a Europe still remains a utopia.

Mainstream SPD spokespersons have so far hedged those problems by emphasizing not substantial progress toward defensive restructuring at home, but the need for mutual restructuring and, especially, the "political civilization process" that supposedly comes with an East–West debate about a nonprovocative defense. Karsten Voigt postulates that "the debate itself already changes attitudes and perceptions; it civilizes." He goes on to stress "development of common criteria for force planning" and "communication and changing of threat perceptions."[54]

Only the minority of determined proponents of alternative defense in the SPD, such as Andreas von Bülow, perceive a potential for unilateral Western restructuring. Reasoning that an area defense with the strong sword elements of the Bülow proposal would guarantee a successful forward defense and damage limitation, they even believe that restructuring would for the first time in history make NATO defenses steadfast enough to justify the removal of all nuclear weapons from West German territory. In this way they rationalize their call for unilateral denuclearization.[55] The SPD as a whole, on the other hand, is only committed to *negotiated* zero-solutions for battlefield and short-range nuclear forces (e.g., in the form of "nuclear-free corridors") and also rejects deep unilateral reductions in heavy weapons.[56] "Owing to her strong capabilities near the border, the Soviet Union currently has more leeway for unilateral steps in Central Europe than NATO. A truly decisive step would be the reform of Warsaw Pact strategy from an offensive into a forward defense. This presupposes a change in the doctrine and structure of Warsaw Pact forces, which, however, would then also enlarge NATO's latitude for unilateral measures."[57] Consequently, the official SPD outlook on the concept's chances of success is quite sober: "To achieve this on both sides is an ambitious long-term goal which can be realized step by step only."[58]

While hopeless in the short and medium terms as a step toward the replacement of deterrence, structural incapability of attack has more realistic implications for conventional arms control. The authors of that concept neither impute aggressive schemes to NATO nor claim that it is even prepared to conduct a strategic offensive, while admitting that the Warsaw Pact is so prepared. But they think that the Soviet Union could not be blamed for perceiving an offensive potential in NATO's tank-heavy and, through FOFA and AirLand Battle, increasingly forward-oriented forces. They are structurally—if not in peacetime numerically—capable of attack. If both sides agree gradually to realign their force structures in such a way that they end up being *structurally* incapable of threatening each other, then a process of mutual arms reductions at the conventional level becomes possible in two ways. First, during the process of structural realignment, light defensive equipment such as antitank and antiaircraft missiles would progressively replace heavy offensive equipment such as tanks, heavy artillery, attack aircraft, and attack helicopters. This would represent arms reduction at the materiel if not the personnel level. Disengagement schemes, such as corridors free of heavy equipment, and various confidence-building measures would also play a role at this stage. If this process remained mutual throughout, the progressive decline of the Warsaw Pact's offensive capability would offset NATO's progressive loss of armor and heavy firepower, while the sine qua non of a continuous forward defense of the Central Front would remain unquestionable. Once the East had reached the juncture of offensive harmlessness in its part of the restructuring, that sine qua non could be dropped, and the second stage of numerical reductions could get underway.[59]

As the principle for a solution of NATO's arms control–security dilemma—the contradiction between the need not to fall short of today's minimum defense of the Central Front and the need to reach mutual reductions in forces, however asymmetrical—structural incapability of attack is logically compelling if the Soviet Union goes along as envisioned. It solves the impossibility of sizeable numerical reductions by inserting into the arms control process a stage of mutual defensive restructuring, after which that impossibility would no longer hold true. It enables the West to continue conventional negotiations with an attractive offer, enticing the Soviet Union to make even more substantial concessions on elements of her offensive superiority.[60] It would gradually transform forward defense into a forward defensive belt composed of light units, and thus it may increasingly resemble the area defense condemned previously. However, structural incapability of attack does not necessarily entail the dismantling of linkages to strategic levels of deterrence. It need not imply denuclearization or unreasonable constraints on Allied forces, although some of its advocates usually mention it in one breath with those counterproductive ideas.[61] As to the operational–tactical and logistical shortcomings of light forces, the gradual, *mutual,* and strategically backed process of conversion to a light format should minimize the risks these shortcomings would evoke during the transition phase of a not yet eliminated offensive potential of Soviet forces.

As currently formulated by the SPD, however, that principle is still a far cry from qualifying as a guide for realistic action. For one thing, it is still too mixed up with those hazardous ideas of unilaterally adopting area defense and denuclearizing West Germany. It also has not been well thought through at the tactical and resource levels. As some authors admit, a *Bundeswehr* in the process of defensive restructuring will have to raise its manpower input in order to obtain a constant defensive output, because light infantry units are more personnel-intensive than armored units.[62] This means even more reliance on reserves than already planned today in expectation of the manpower crunch from low birthrates in the 1970s. In fact, the West German army's combat forces would for the most part have to change to a cadre or militia structure.[63] Quite apart from the dubious value of reservists in a light infantry role,[64] a structure that depends upon large-scale mobilization to reach its defensive effectiveness would violate the strategic imperative of crisis stability. The need to mobilize in response to every ambiguous action by hostile forces in order to preclude defeat through surprise attack aggravates the crisis, dulls the alertness, and exhausts the economy of the defender and deprives him of diplomatic flexibility. Thus, during the transition period of a residual Soviet offensive threat, a concept of defensive restructuring based on the massive use of reserves must be complemented by some mechanism that facilitates a calm and informed decision for mobilization. Tactically, it is questionable whether defensively structured units should necessarily renounce the protection of *armor* in addition to the use of *battle tanks*. The most noteworthy defensive advantages of the present force posture would be lost (i.e., operational mobility for quick defensive concentrations, tactical mobility under indirect fire, some degree of protection against nuclear and chemical attack, and the salutary effect of armored protection, mobility, and crew spirit on morale). And, as outlined in the previous section on alternative defense, no defensive advantages would be gained. It is beyond the scope of this discussion to suggest a solution to those problems associated with the mutual conversion to force postures more structurally incapable of attack.[65] However, it is clear that nonprovocative defense represents as yet an immature concept even for limited purposes of arms control.

Because of the described definitional, political, and substantive problems, structural incapability of attack will have a negligible impact on the evolution of Western strategy and East–West arms control for the next ten to twenty years.

NOTES AND REFERENCES

1. See Rüdiger Moniac, "Sichtbare Anti-Atomwaffenströmung," *Die Welt,* June 15, 1988, p. 4.
2. Alfred Dregger, "Disarmament with Security: A German View of Current Alliance Developments," *The Atlantic Community Quarterly* (Winter 1987–1988), p. 410 (emphasis in text).
3. See "Der deutsche Vorschlag für konventionelle Stabilität, "*Europäische Wehrkunde/WWR* 4 (1988), p. 217.
4. See "'Bundeswehrplan 1989' und 'Heeresstruktur 2000'; weniger Menschen und weniger Mittel," *Europäische Wehrkunde/WWR,* 3 (1988), p. 119.
5. See Helga Haftendorn, "Transatlantische Dissonanzen; der Bericht Über 'Selektive Abschreckung' und die Strategiediskussion in den USA," *Europa-Archiv* 8 (1988), pp. 218–19.
6. See Commission on Integrated Long-Term Strategy, *Discriminate Deterrence* (Washington, D.C.: January 12, 1988).
7. See Peter Schmidt, *Europeanization of Defense: Prospects of Consensus?* (Santa Monica: The Rand Corporation, December 1984), pp. 26–28.
8. Ibid., pp. 4–5.
9. See "Empfehlungen der Versammlung der Westeuropäischen Union während des zweiten Teils ihrer 33. Sitzungsperiode in Paris vom 30. November bis zum 3. Dezember (Auswahl)," *Europa-Archiv* 4 (1988), p. D103.
10. See WELT-Gespräch with the French foreign minister Giraud, "Die Abrüstung sollte bei den Grossen beginnen," *Die Welt,* May 21, 1987, p. 4; "Das Bündnis braucht zwei starke Pfeiler" (interview with Foreign Secretary Geoffrey Howe), *Die Welt,* May 9, 1987, p. 5; "Die WEU will sich Zeit nehmen," *Süddeutsche Zeitung,* April 29, 1987, p. 7; "Dregger: 'Es kommt vor allem auf uns an'" (interview with Alfred Dregger), *Die Welt,* May 15, 1987, p. 5; "Niederlande wünschen Strategie-Debatte," *Süddeutsche Zeitung,* January 14, 1988, p. 9.
11. "Das Bündnis braucht zwei starke Pfeiler."
12. Werner Weidenfeld, "Europas Sicherheitsstrukturen müssen neu organisiert werden," *Die Welt,* May 19, 1987, p. 6.
13. See Samuel F. Wells, Jr., "The United States and European Defence Co-operation," *Survival* (July–August 1985), pp. 165–66.
14. Rüdiger Moniac, "Jetzt taucht die Europäische Atomstreitmacht wieder auf," *Die Welt,* May 5, 1987, p. 2.
15. See Wells, "The United States."
16. See "European Security Co-operation," *Survival* (July–August 1987), p. 376.
17. Ibid.
18. See, for a recent expression of this proposal, "London und Paris arbeiten an gemeinsamer Träger-waffe," *Die Welt,* February 18, 1988, p. 6.
19. See "Militärische Kooperation Paris–London," *Die Welt,* February 1, 1988, p. 6.
20. Ibid.
21. See President Francois Mitterand, "'Ihr Deutschen seid ein grosses Volk. Bewahrt Eure Identität!'" (interview), *Die Welt,* January 18, 1988, p. 7; Christian Millotat and Hartmut Bühl, "Kooperation

zwischen Paris und Bonn; Partnerschaft zum Nutzen der Allianz," *Europäische Wehrkunde/WWR* 2 (1988), pp. 70–74.

22. Franz H. U. Borkenhagen, "Der europäische Pfeiler der westlichen Allianz," *Aus Politik und Zeitgeschichte* B18 (1988), p. 37.
23. "Kommuniqué über die Ministertagung der Euro-Gruppe in der NATO am 30. November 1987 in Brüssel," *Europa-Archiv* 4 (1988), p. B102.
24. Schmidt, *Europeanization of Defense,* p. 27.
25. See Wells, "The United States," p. 161; Millotat and Büshl, "Kooperation zwischen Paris und Bonn."
26. See Peter Ruge, "Die Franzosen wollen weiter die massive Abschreckung," *Die Welt,* March 5, 1988, p. 2.
27. See "Andreotti: Militärintegration Bonn–Paris ein Irrweg," *Süddeutsche Zeitung,* January 18, 1988, p. 7.
28. See "Kommuniqué der Ministertagung der Unabhängigen Europäischen Programmgruppe (IEPG) in Sevilla am 22. Juni 1987," *Europa-Archiv* 4 (1988), pp. D92–D93.
29. See Samuel P. Huntington, "Conventional Deterrence and Conventional Retaliation in Europe," *International Security* (Winter 1983–1984), pp. 32–56.
30. Robert Jervis, "Cooperation under the Security Dilemma," *World Politics* (January 1978), pp. 167–214.
31. Dieter S. Lutz, "Was heisst 'Strukturelle Nichtangriffsfähigkeit' (i.e.S.)? Versuch einer Annäherung," *S–F (Sicherheit und Frieden)* 1 (1987), p. 2.
32. See Heinz Brill, ed., *Bogislaw von Bonin—Opposition gegen Adenauers Sicherheitspolitik: eine Dokumentation* (Hamburg: Verlag Neue Politik, 1976).
33. Horst Afheldt, *Verteidigung und Frieden: Politik mit militärischen Mitteln* (Munich and Vienna: Carl Hanser Verlag, 1976); see also Carl Friedrich von Weizsäcker, *Wege in der Gefahr* (Munich and Vienna: Carl Hanser Verlag, 1976).
34. See Horst Mendershausen, *Territorial Defense in NATO and Non-NATO Europe* (Santa Monica: The Rand Corporation, February 1973); Adam Roberts, *Nations in Arms: the Theory and Practice of Territorial Defense* (New York: Praeger Publishers, 1975); E. W. Paxson, M.G. Weiner, and R. A. Wise, *Interactions between Tactics and Technology in Ground Warfare* (Santa Monica: The Rand Corporation, January 1979); Steven L. Canby, "Territorial Defense in Central Europe," *Armed Forces & Society* (Fall 1980), pp. 51–67.
35. Emil Spannocchi, "Verteidigung ohne Selbstzerstörung," in *Verteidigung ohne Schlacht* (Munich and Vienna: Carl Hanser Verlag, 1976).
36. Guy Brossollet, "Das Ende der Schlacht," in *Verteidigung ohne Schlacht* (Munich and Vienna: Carl Hanser Verlag, 1976).
37. Franz Uhle-Wettler, *Gefechtsfeld Mitteleuropa: Gefahr der Übertechnisierung von Streitkräften* (Munich: Bernard & Graefe Verlag, 1980).
38. See "Die Struktur der Bundeswehr," *Der Spiegel* 25 (1980), p. 52, "Spiegel–Gespräch: 'Verteidigt wird vor Hamburg und vor Kassel'," *Der Spiegel* 26 (1980), pp. 23–26; "' Ein Heer für den Angriff'," *Der Spiegel* 21 (1982), pp. 80–103.
39. Peter Bender quoted in Hans-Gert Pöttering, "Eine neue europäische Sicherheitspolitik?," *Aussenpolitik* 2 (1985), p. 149.
40. Hans-Joachim Löser quoted in Ludwig Schulte, "Das Schlagwort von der 'defensiven Umrüstung'," *Europäische Wehrkunde/WWR* 9 (1985), p. 122.
41. Karsten D. Voigt, "Konventionelle Stabilisierung und strukturelle Nichtangriffsfähigkeit," *Aus Politik und Zeitgeschichte* B18 (1988), p. 24.
42. Horst Afheldt, *Defensive Verteidigung* (Hamburg-Reinbek: Rowohlt-Verlagsgesellschaft, 1985); Günter Vollmer, "Zur Sicherheitspolitik und Landesverteidigung der Bundesrepublik Deutschland," in Ingo Günther and Günter Vollmer, *Verteidigung statt Vernichtung: Wege aus der atomaren Konfrontation* (Munich: ibf-Verlag, 1983), pp. 80–135.
43. Norbert Hannig, "The Defense of Western Europe with Conventional Weapons," *International Defense Review* (1981), pp. 1439–43; same author, *Abschreckung durch konventionelle Waffen: das*

David–Goliath–Prinzip (Berlin: Berlin Verlag Arno Spitz, 1984); Johannes Gerber, "Die konventionelle Nichtangriffsfähigkeit als militärische Dimension der Sicherheit in Mittel—und Nordeuropa," *S–F (Sicherheit und Frieden)* 1 (1987), pp. 23–30.

44. Studiengruppe Alternative Sicherheitspolitik (SAS), "Landstreitkräfte zur Verteidigung der Bundesrepublik Deutschland" and Lutz Unterseher, "Für eine tragfähige Verteidigung der Bundesrepublik Deutschland: Grundgedanken und Orientierungen," in SAS, ed., *Strukturwandel der Verteidigung—Entwürfe für eine konsequente Defensive* (Cologne: Westdeutscher Verlag, 1984).
45. Andreas von Bülow, "Vorschlag für eine Bundeswehrstruktur der 90er Jahre: Einstieg in die strukturelle Nichtangriffsfähigkeit," *Europäische Wehrkunde/WWR* 11 (1986), pp. 636–45; Hans-Joachim Löser, *Weder rot noch tot: Uberleben ohne Atomkrieg: eine sicherheitspolitische Alternative* (Munich: Olzog Verlag, 1981); same author, "Strukturelle Nichtangriffsfähigkeit," *S–F (Sicherheit und Frieden)* 1 (1987), pp. 51–60; Albrecht C. von Müler, *Die Kunst des Friedens: Grundzüge einer europäischen Sicherheitspolitik für die 80er und 90er Jahre* (Munich: 1984); Lothar Penz, "Strategische Bedingungen alternativer Verteidigung," *Junges Forum* (Spring 1984), p. 37ff.
46. See Vollmer, "Zur Sicherheitspolitik," p. 119.
47. Lutz Unterseher quoted in David Gates, "Area Defence Concepts: the West German Debate," *Survival* (July–August 1987), p. 307.
48. John C. Slessor, *Strategy for the West* (New York: William Morrow and Company, 1954); B. H. Liddell Hart, "Shield Forces for NATO," *Survival* (May–June 1960): pp. 108–11; Ferdinand O. Miksche, *The Failure of Atomic Strategy: a New Proposal for the Defense of the West* (New York: Praeger Publishers, 1959).
49. For the contrary view, see Philipp Borinski, "Another Look at USAREUR Deployment," *Military Review* (March 1987), pp. 49–61.
50. See Siegbert Rietzler, "Erfahrungen aus Kleinkrieg und Jagdkampf," *Truppendienst* 2 (1979), pp. 155–57.
51. See Christian Krause, "Strukturelle Nichtangriffsfähigkeit im Rahmen europäischer Entspannungspolitik," *S–F (Sicherheit und Frieden)* 3 (1987), p. 189.
52. Lutz, "Was heisst 'Strukturelle Nichtangriffsfähigkeit'," p. 2.
53. See EWK, "Der deutsche Vorschlag für konventionelle Stabilität," *Europäische Wehrkunde/WWR* 4 (1988), p. 217.
54. Voigt, "Konventionelle Stabilisierung," p. 21.
55. See Andreas von Bülow and Helmut Funk, "Zum Problem der Strukturellen Nichtangriffsfähigkeit," *S–F (Sicherheit und Frieden)* 1 (1987), p. 15.
56. "Dokumentation: Beschlüsse des SPD-Parteitages; Politik der doppelten Partnerschaft," *Europäische Wehrkunde/WWR* 10 (1986), pp. 595–97.
57. Voigt, "Konventionelle Stabilisierung," p. 34.
58. Karsten D. Voigt, "Von der nuklearen zur konventionellen Abrüstung in Europa—Kriterien konventioneller Stabilität und Möglichkeiten der Rüstungskontrolle," *Europa-Archiv* 14 (1967), p. 418.
59. See Bülow, "Vorschlag für eine Bundeswehrstruktur"; Bülow and Funk, "Zum Problem der Strukturellen"; Müller, *Die Kunst des Friedens;* Voigt, "Von der nuklearen zur konventionellen Abrüstung"; Egon Bahr, "Korridor ohne Waffen für den Angriff," *Europäische Wehrkunde/WWR* 3 (1987), pp. 140–43; Kurt Kister, "Dominanz der Defensive," *Süddeutsche Zeitung,* February 5, 1988, p. 11; "Nur noch rückwärts," *Der Spiegel* 43 (1987), pp. 38–40.
60. See EWK, "Der deutsche Vorschlag."
61. See Bahr, "Korridor ohne Waffen"; Bülow and Funk, "Zum Problem der Strukturellen"; Oskar Lafontaine, *Angst vor den Freunden* (Hamburg-Reinbek: Rohwolt-Verlagsgesellschaft, 1983).
62. See Bülow, "Vorschlag für eine Bundeswehrstruktur"; Löser, *Weder rot noch tot;* also Uhle-Wettler, *Gefechtsfeld Mitteleuropa.*
63. See Bülow, "Vorschlag für eine Bundeswehrstruktur"; Löser, *Weder rot noch tot.*
64. See Siegfried Thielbeer, "Kaderung und grössere Abhängigkeit von Reservisten bei der Bundeswehr," *Frankfurter Allgemeine Zeitung,* April 5, 1988, p. 4.
65. For a possible solution, see Philipp Borinski, "Mitigating West Germany's Strategic Dilemmas," *Armed Forces and Society* (August 1989), pp. 531–49.

3

Future Requirements for Theater Weapons in Western Europe

INTRODUCTION

What are the likely future roles of and requirements for theater nuclear weapons[1] in light of changing strategic concepts and evolving political, economic, military, and technological factors? Which weapons characteristics should be developed to meet future requirements in light of current assessments of these future military requirements and likely constraints on the development and deployment of nuclear weapons, particularly in Europe?

Theater nuclear weapons have long played a key role in NATO's plans to deter Soviet–Warsaw Pact aggression and to defend Western Europe if deterrence fails. These weapons (along with other elements of the U.S. nuclear arsenal) are also an important symbol of the U.S. commitment to protect its NATO partners, thereby assuring that U.S. and Western European security are linked. However, in recent years the utility of theater nuclear weapons in these roles has been increasingly challenged on both sides of the Atlantic, raising serious questions about what their role will be in the future, and thus what requirements there will be for modernization of the current stockpile.

The utility of theater nuclear forces is under heightened scrutiny today as result of dramatic political changes in Eastern Europe and in the aftermath of the December 1987 Intermediate-Range Nuclear Forces (INF) Treaty signed by President Ronald Reagan and Soviet General Secretary Mikhail Gorbachev, mandating the elimination of all intermediate- and shorter-range nuclear forces (those with ranges between 300 and 3,000 miles) from Europe. The treaty drastically cuts NATO's overall theater nuclear capacity in the region, leaving a small number of nuclear maritime depth charges (range: 0), hundreds of artillery-fired atomic projectiles (AFAPs; range, about 9 miles), approximately eighty aging *Lance* missiles (range, about 66 miles), and hundreds of nuclear bombs for NATO's dual-capable aircraft (range 180–200 miles). All told, there are approximately 4,500 warheads remaining in the European theater.

The treaty has also increased calls within Europe, and particularly within West Germany, to negotiate further reductions in theater nuclear forces. While reductions are usually linked by Supreme Allied Commander, Europe (SACEUR) to modernization of the remaining theater nuclear forces, history suggests that reductions will occur even if modernization does not. So far, most European governments, with the exception of the Federal Republic of Germany, have resisted further reductions, but the recent developments in Eastern Europe which bring the prospect of lowered political tensions and major reductions in conventional forces, could significantly affect the direction of this debate.

Also leading the European allies—and many American legislators—to question the utility of the modernization of theater nuclear weapons is Mikhail Gorbachev's "peace offensive" aimed at Western Europe. Gorbachev has made a number of concrete military and political proposals addressing not only the remaining short-range nuclear weapons but also the East–West conventional imbalance in the theater. The dramatic announcement in his United Nations speech of December 7, 1988, of unilateral reductions in Soviet conventional capabilities have led to an acceleration of the Conventional Forces in Europe (CFE) negotiations. There also have been calls by the Europeans and some Americans for reciprocal U.S. and NATO unilateral action. Despite their popular reception, some see these moves as a calculated attempt to create a schism between the United States and the allies; others see this peace offensive as a genuine effort on the part of Gorbachev aimed at improving political and economic relations between the Europeans and the Soviets, a step essential to Gorbachev's efforts to reform the Soviet economy, *perestroika*.

All in all, the problems facing the alliance can be summed up as: should we modernize? what should we modernize? and how soon should we do it? This last question is especially important. There now is a strong temptation to take a "wait-and-see" attitude with respect to Gorbachev and future alliance trends, but there are also sound reasons for proceeding with modernization as rapidly as possible. In any case, the long lead times some new systems require before they would be operable makes a wait-and-see approach partly unavoidable anyway, at least with respect to deployment. Thus, we focus on both long-term and short-term issues, but we are particularly concerned about the prospects for implementing the immediate modernization recommendations of the present SACEUR, Supreme Allied Commander, Europe, General John Galvin. To integrate the analysis, three alternative political–military futures, each of which would impact differently on nuclear weapons requirements, are examined. For each future setting we identify weapons requirements and characteristics that seem to be consistent with that future setting. The three scenarios we examine are categorized as, first, a continuation of current trends; second, reduced East–West tensions and an eroding NATO alliance; and, third, increased tensions and a strengthening of the NATO alliance ties.

SETTING THE CONTEXT: FACTORS AFFECTING THEATER NUCLEAR FORCES

NATO's Nuclear Dilemmas

Any examination of the future role of nuclear weapons in NATO must take into account some traditional differences between the United States and Europe with respect to that role.[4] Generalizations are dangerous, for there is no unanimity of views among experts or among the general public in either the United States or Europe, and views about the role of nuclear weapons are changing on both sides of the Atlantic. It is thus not correct to say that there is an "American view" and a "European view" on issues related to nuclear weapons. But there are recognizable differences between most European experts and most American experts on the role of nuclear weapons in NATO strategy and thus in the way these expert views are reflected in public opinion and political debate.

These differences are not surprising, given the different geopolitical positions of the European and North American States. In general, the people of Western Europe share a malaise at being heavily dependent on a distant power—the United States—for their security, and they frequently express doubts about the steadfastness of the U.S. commitment to use nuclear weapons in their defense. At the same time many Europeans are increasingly uneasy with the presence of large numbers of nuclear weapons on their soil. If European views about nuclear weapons sometimes appear ambivalent or even inconsistent, this is an accurate reflection of reality.

Traditionally, the United States, while recognizing the vital importance of nuclear weapons to deterrence, has sought to develop options to sole or heavy reliance on nuclear weapons, and particularly on the threat of massive retaliation, as the cornerstone of deterrence. The United States introduced *massive retaliation* in the 1950s, when we enjoyed a nuclear dominance, and most of our allies embraced the concept with enthusiasm. However, the United States has spent much of the past twenty-five years trying to disengage NATO strategy and force posture from that policy. The effort began in earnest in the early 1960s under Robert McNamara, and it continues to this day with the proposals of the recent SACEUR, General Bernard Rogers, to reduce NATO's heavy dependence on nuclear weapons and modernize their conventional as well as their nuclear forces.

To a large degree this disengagement from massive retaliation reflects a long-held apprehension on the part of U.S. leaders about being drawn prematurely—or possibly even against their will—into an escalating crisis abroad. Of course, the United States has never sought to eliminate the important role of nuclear weapons for deterrence. Rather, U.S. leaders have emphasized the importance of flexibility—to have options for conventional operations, as well as for the early use of nuclear weapons and, if they are used, to have options to do so in a limited fashion. The concept of limited use of nuclear weapons in the defense of Europe is challenged by many American experts,[5] but it remains part of official NATO policy (e.g., in the form of Selective Employment Plans).

Europeans, in general, have been reluctant to make a concerted effort aimed at building up NATO's conventional forces, and they have also looked warily at American efforts to incorporate additional flexibility into NATO war plans. Flexible response, NATO's official war-fighting doctrine, requires selective nuclear options below massive levels. Nonetheless, many Europeans see measures to limit nuclear use or control escalation as illusory. Indeed, deliberately posing the risk of escalation is an important element of NATO's doctrine for deterrence. Many European defense experts and political leaders recognize that even if the military imbalance in Europe is reduced as a result of CFE, nuclear weapons still will be needed for many years. They nonetheless prefer to ignore the need for modernization.

The reasons for this attitude are not hard to appreciate. "Fixing" NATO's defense posture to reduce the alliance's reliance on nuclear weapons and increase its conventional capabilities, in light of a militarily formidable Soviet threat (even if, and after, Gorbachev's unilateral conventional cuts are fully completed), is an expensive proposition that would force all NATO governments to confront major dilemmas in fiscal priorities. Recognizing the impact of such changes in peacetime, the Europeans have traditionally deferred major modernization of theater nuclear forces and have consistently fallen short of their conventional force goals.

Even more fundamentally, few Europeans are convinced today that a serious military threat remains. Nor are they persuaded that there really is a better choice than a strategy that relies heavily on massive retaliation. Surely, the "limited" use of nuclear weapons on the battlefield has no appeal to Europeans whose territory would constitute the battlefield. To most Europeans this is not a nice, clean tactical nuclear war, but a one-sided strategic engagement in which their former major adversary, the Soviet Union, remains a sanctuary and their major ally, the United States, avoids the risks being imposed on Europe.

While U.S. leaders like to envisage theater nuclear forces as a firebreak on the escalation ladder, many of their European counterparts see the employment of these weapons as the fuse that would ignite the exchange between U.S. and Soviet strategic nuclear forces. These observers see the prospect of strategic engagement as the most effective means of deterring a potential Soviet–Warsaw Pact aggression, and they believe that the risks and uncertainties of uncontrolled escalation are the core of deterrence.

To further complicate matters, there are now many Europeans, particularly outside of government, who reject the deterrent effect of coupling U.S. strategic and theater forces and would simply prefer that the latter be completely removed from European soil. Many of these observers perceive nuclear weapons to be a greater threat than the Soviets, and they question whether "deterrence" as a strategy is actually necessary. These observers see *any* theater nuclear weapons as dangerously increasing the risk of escalation and would like to see a follow-on agreement to the INF Treaty, rather than a concerted Western effort to modernize the remaining battlefield nuclear weapons.

Another traditional area of dispute, closely linked to the above set of concerns, is defense burden-sharing. Historically, NATO has looked to the United States—its largest and most militarily capable member—to carry the bulk of the defense burden. Yet, with the expansion of U.S. global commitments over the past three decades and the full-scale recovery of the Western European states during the same period, the United States has pressed the Europeans to spend more on defense[6] to improve the equipment, readiness, and wartime sustainability of NATO's conventional forces. But conventional force improvements are costly, especially when compared to nuclear force modernization. Hence, there is an almost self-perpetuating tendency on the part of the allies to put off such improvements (thereby avoiding major budget increases) and to continue to rely heavily on nuclear weapons. This predilection, however, does not necessarily generate support for measures to modernize these weapons. The modernization of these weapons also will impose costs, particularly for the United States, and thus the issue of how these costs should be allocated also has become part of the burden-sharing debate.

From the U.S. perspective, the inability or unwillingness of the allies to devote a greater share of resources to defense—to make flexible response more "flexible"—is a source of continuing frustration, notably in Congress. A highly visible manifestation of this frustration—and one that not only reveals but nurtures the seeds of isolationist sentiment—has been the proposals advanced over the years, particularly by Senator Sam Nunn, to withdraw U.S. troops from Europe unless the NATO allies devote more resources to the common defense effort. Another Western ally, Japan, has often been the focal point of similar criticisms.

One such proposal, offered in 1984 by Senator Nunn, was defeated by only a handful of votes. An attempt by Representative Patricia Schroeder to attach a similar measure to the FY87 defense authorization bill was overwhelmingly rejected by the House of Representatives. However, a proposal in 1989 to make cuts in manpower associated with the elimination of INF systems was approved by Congress. Should the reduced political tensions in Europe and the tight budget environment continue in the next few years, and a majority of the key allies continue to spend below the agreed NATO goal of 3 percent per year increase in defense spending, let alone the 4 percent level of growth which the former SACEUR determined was necessary, support for troop withdrawal initiatives will surely grow.[7]

A related area where some movement could occur, precisely because of the tight budget situation, is in efforts to foster cooperation among the allies on production and procurement of weapons. It has long been recognized that NATO's weapon-development procedures, which are, in general, poorly coordinated and largely inefficient, contribute to frictions over defense spending and burden-sharing and, more important, hamper the integration of combat forces on the battlefield. The rhetorical support for such cooperation has not been matched with commensurate achievements. This primarily results from competing national economic priorities and concerns over technology transfers to the Soviets. These concerns have not completely abated. But now, in the face of the large U.S. federal budget deficit, the strictures imposed on defense spending by the Gramm–Rudman budget-balancing law, and the constraints imposed by budget compromises hammered out by Congress and the administration, there is a growing realization that it will be essential to make the most efficient use of limited future resources.

Toward this end, Congress in 1985 authorized $200 million for joint U.S.–

NATO research projects. The program was favorably received by both Congress and European governments, and some progress has already been reported. However, it is uncertain whether future projects related to nuclear weapons—including the largest and perhaps most controversial, the Strategic Defense Initiative (SDI)—can be moved forward on a cooperative basis. Indeed, the most promising prospects for joint research and development efforts are in the conventional realm.

Another traditional factor that has been an ongoing source of controversy within the alliance is the nature and extent of the Soviet threat. When NATO was founded, the Western allies enjoyed a distinct technological superiority in weaponry over the Warsaw Pact, offsetting the latter's superiority in manpower. At that time, too, there was a virtual consensus on the threat posed by the USSR. Since then, and especially over the past fifteen years, the Soviet Union and its allies have greatly improved the quality of their weapons and equipment to the point where many experts became convinced that Western qualitative advantages no longer offset Eastern advantages in numbers.

Now, Soviet initiatives to reduce these forces in Europe unilaterally, along with their willingness to accept further asymmetrical reductions in CFE, have added fuel to the debate. There is no longer a strong consensus in the West on the implications of these developments for the security of the alliance. To be sure, it is certainly accurate to conclude, as does former U.S. Ambassador to NATO David Abshire, that "allied political leaders agree that the gap between NATO and the Warsaw Pact is wide."[9] But the problem is not primarily at this level. Instead, it lies in the perceptions of the general public and with many politicians in Europe, who emphasize the lack of Soviet incentives to go to war and the need for NATO (and particularly the United States in its global role) to refrain from actions that might provoke the USSR. In addition, in this view détente and arms control are the preferred path to real security. Karsten Voigt, a prominent military and foreign affairs spokesman for the German SPD, offers a representative view:

> Notwithstanding the Warsaw Pact's enormous military potential, there are no signs the Soviet Union currently poses an acute military threat to Western Europe. It shows no desire to expand militarily into Western Europe. . . . The problem confronting Europe today is less that of continuing Soviet military expansion . . . than of organizing lasting, peaceful East–West coexistence in the face of continuing conflict between the two political systems. . . . Efforts at the peaceful accommodation of interests, the control and reduction of armaments, and military confidence-building measures must receive political priority.[10]

Proponents of this view have been bolstered by the ascension of Mikhail Gorbachev to the top of the Soviet leadership hierarchy and, more recently, by the steps taken toward greater freedom and independence in Eastern Europe which appear to diminish the threat of war. Gorbachev has specifically and repeatedly renounced any intention of taking any hostile actions against Western Europe. He has said that to facilitate *perestroika,* his program to restructure the Soviet economy, he would like to reorient scarce fiscal resources away from the military and into the domestic economy. Such statements are seen by many citizens and politicians in

Western Europe as further evidence of a receding Soviet threat. These views tend to act as a constraint on new deployments and on the peacetime movement of nuclear systems within Europe, a step which could reduce their vulnerability. They also impel political leaders to seek improved security through détente and arms control, rather than through concrete military remedies.

Recent Developments

The INF Treaty

An important recent development affecting theater nuclear forces in Europe is the Treaty for the Elimination of Intermediate- and Short-Range Nuclear Forces, or the INF Treaty. It eliminates Soviet SS-20s, as well as the remaining Soviet SS-3s and 4s, and NATO's *Pershing* II (P-II) ballistic missiles and ground-launched cruise missiles (GLCMs). The Soviets are required to dismantle about 1,800 missiles and the U.S. about 800.

There are a number of positive aspects to the INF Treaty. It removes one form of a particularly grave threat to NATO—the Soviet SS-20s. The treaty also establishes a verification regime which is unprecedented in its detail and comprehensiveness, and for the first time it permits on-site inspections by both U.S. and Soviet monitoring teams. As indicated above, the treaty establishes the principle of asymmetrical reductions, as the Soviets are required to destroy more than twice as many missiles as required by the United States. And perhaps most important, the treaty shows the benefits to be attained from a cohesive alliance, united in its determination to deploy the INF forces in the first instance, and in sticking to its belief that the deployments would bring the Soviets to the arms control negotiating table.

However, there are also some less encouraging aspects of the treaty. It removes NATO's most effective means of reaching targets within the Soviet Union (barring the resort to U.S. strategic systems). While dual-capable aircraft will in some circumstances be able to fulfill this mission, their bases remain highly vulnerable targets, and they have to penetrate sophisticated Soviet air defenses. At the same time, however, the Soviets *retain* their capability to attack West European soil with their own nuclear forces, which could easily be targeted against NATO assets. The treaty could create a "slippery slope" to the denuclearization of Europe. Already it is clear that many West Europeans will increasingly demand a greater emphasis on arms control than on modernization in planning for the European theater, particularly with respect to battlefield nuclear weapons. Many believe that the treaty could lead to a decoupling of Europe from the United States, as some Europeans may no longer find the assurance of a strategic umbrella credible, and still others will no longer find it desirable. The treaty also requires that all GLCMs be banned, thus eliminating the conventional GLCM option.

The INF Treaty could worsen NATO's already tenuous ability to meet the requirements of flexible response. The basic issue is that the INF Treaty narrows certain nuclear options in NATO's strategy. The ability to pose the threat of escalation, including the ability to directly target the Soviet Union, is a central feature of

NATO's strategy. The elimination of the most effective long-range weapons in SACEUR's command could adversely affect deterrence, if the Soviets see NATO's threat to escalate to nuclear weapons as less credible.

The bottom line, however, is that the INF Treaty is a fact of life. It is ratified and many of the missiles have almost all been destroyed. It is tempting to spend time—depending upon one's perspective—bemoaning or cheering the loss of this capability. It is more important, however, to consider what is best for NATO in light of the treaty and how NATO can best meet its military and political objectives in the changed circumstances. Whatever the benefits or drawbacks to the treaty, the fact that a powerful part of NATO's arsenal no longer exists must be accepted and accounted for in future planning. Modernization of the remaining nuclear forces in Europe would help to offset the loss of the INF forces, but there are no guarantees that this can be accomplished in the current political environment. There are strong forces in the FRG opposed to the modernization of theater nuclear forces, and if the United States were to press too hard for modernization now, it could create a political crisis of major proportions in the alliance; but without strong U.S. leadership, modernization is unlikely to occur. How to handle the modernization of these forces represents a major policy dilemma for the Bush administration.

The Changing Threat

An issue which has come into focus in the last few years is the changing nature of the Soviet threat. The Soviet proposals to reduce their forces in Europe have had a dramatic impact on public attitudes in the West and, if carried through, will significantly alter the military balance in Europe. Furthermore, Gorbachev's toned-down rhetoric has resulted in an increasingly benign perception of the Soviet Union held by growing numbers of Americans and Europeans. Nonetheless, there have been significant changes in Soviet doctrine and force structure, and the West cannot fail to address them.

For example, five Soviet tactical aircraft (the SU-7, MIG-21, MIG-27, SU-17, and SU-24) are considered to be nuclear capable, and approximately 1,200 of such aircraft are deployed in peacetime, where they constitute a direct threat to NATO Europe. The Warsaw Pact also has deployed nuclear-capable artillery in recent years, which have greater range than comparable NATO systems. And even after the elimination of INF, the Soviets will retain hundreds of short-range missiles capable of hitting most military targets in Western Europe. And even if CFE succeeds in eliminating some of the major disparities between NATO and WTO forces, the Soviets will retain very formidable military forces in Europe and a large nuclear arsenal.

Perhaps even more significant than these force improvements are the recent changes in Soviet doctrine with respect to nuclear-weapon employment. Divining Soviet doctrine is always difficult, for they do not go out of their way to make that doctrine clear to the West; certainly some deception is involved in their public doctrinal statements. Still, the Soviets have spoken extensively about developing a more defensive doctrine and also about exploiting advanced technology more effectively even if forces are reduced. New evidence suggests that they would try to avoid the early use of nuclear weapons in a European war, relying on those weapons primarily to deter NATO's first use. They have attempted to develop a capability and doctrine that would permit them to achieve their military objectives without recourse to nuclear weapons. There is also evidence to suggest they might use nuclear weapons selectively in the theater if and when they are used.[11]

It is likely that, if there were to be a nonnuclear phase at the outset of a conflict, the Pact would give high priority to attacking NATO's highly vulnerable nuclear forces during that phase. The Soviets are steadily acquiring the capabilities to locate, identify, and attack high-value NATO targets, such as remaining nuclear weapon deployments and nuclear storage sites, from long range with nonnuclear weapons. Nuclear forces and facilities are also likely to be high-priority targets for the initial waves of aircraft attack.

Rapidly moving division-sized operational maneuver groups (OMGs) could pose a significant threat to forward-deployed nuclear weapons, such as nuclear artillery. Operational maneuver groups have been designed to probe NATO weak spots swiftly and to penetrate deep into allied territory in order to place a strong force behind the allies' main defense belt. Such penetration maximizes the potential for disrupting NATO defense efforts, for example, by cutting critical communication links and by fostering chaos and confusion. Toward this end, OMGs would seek out key targets, such as command and control centers, nuclear weapons, and nuclear storage sites. A primary objective would be rapidly to limit or to eliminate NATO's potential for nuclear response. Such response, it is assumed, will take time, given the likely delay in gaining authorization for nuclear release from political authorities. Operational maneuver groups seek to exploit this time lag, thus helping to assure that a conflict can be kept at the conventional level. If successful, this rapid effort would contribute to the Soviet's important military objectives of seizing adjacent territory intact, of preventing the West from mobilizing its superior economic and industrial resources, and from exploiting potential political divisions between the Soviets and the Eastern Europeans.[12]

As a result of these apparent changes in Soviet doctrine and capability, NATO faces a set of dilemmas on how to deal with nuclear weapons in a crisis or nonnuclear attack. Unless nuclear weapons are dispersed rapidly from their storage sites and the delivery systems from their peacetime locations, NATO's nuclear forces will be highly vulnerable to attack even, as we have suggested, in a possible nonnuclear phase of conflict. But dispersal inevitably provides the enemy with signals that could invite preemption, and for this very reason a decision to disperse nuclear forces could well be delayed to avoid exacerbating a crisis. Even when forces and weapons are deployed to the field, they will be vulnerable to being overrun in forward areas. This pressure to "use them or lose them" is exacerbated by the fact that nuclear weapons are concentrated in relatively few storage sites in peacetime because proliferation is costly. Moreover, dispersal is almost never exercised in peacetime due to the costs involved and to the political sensitivity of conducting realistic exercises involving the movement of nuclear weapons in the densely populated areas of Europe. As a consequence, the enemy can target relatively few high-value targets, and he is likely to obtain very clear signals when a movement of weapons occurs.

Changing Public Attitudes

One new development that constrains NATO's efforts to deal effectively with the above dilemmas is the growing public and political opposition to nuclear weapons. We have noted that many political leaders and publics see nuclear weapons as an increasingly risky basis for deterrence, and they are skeptical and fearful of the war-fighting role assigned to them. These concerns have long permeated public debates (and are buttressed in many nations, like the FRG, by a more general antinuclear sentiment directed against nuclear energy). The opposition of Europeans to the acceptance of enhanced-radiation weapons in the late 1970s is a well-known example of the influence of mass opinion on NATO decision making. The Dutch government, in an attempt to make more tolerable the INF deployments to which it had agreed in the face of unrelenting public and political opposition, announced in November 1985 that in exchange for accepting the deployment of cruise missiles in the Netherlands it was necessary to relinquish its obligation to NATO to carry out nuclear fighter-bomber and antisubmarine combat roles.

To complicate matters further, the Soviets have strived—through the use of Gorbachev's "peace offensive" and with threats and warnings about the fate of détente and arms control—to exploit public concern and unease with nuclear weapons. These actions could potentially divide NATO and undermine the confidence of the allies in their security measures. While the Soviet anti-INF campaign, launched in the early 1980s to prevent NATO from deploying the P-IIs and the GLCMs, was unsuccessful, they have worked hard and with skill to foster anti-SDI opinions over the last six years. Partly as a result of these pressures, NATO political leaders are increasingly hesitant to pursue steps that may be seen as enhancing NATO's dependence on nuclear weapons in any way.

Yet political opposition has not paralyzed all NATO actions. The decision to deploy intermediate-range nuclear forces was an important example of NATO's ability to act in solidarity. This decision was more significant from a political standpoint than from a military one. The P-IIs and GLCMs unquestionably added an important military dimension to NATO defense, but the solidarity of NATO in standing behind the decision, in the face of very strong Soviet pressures, was even more significant to the health of the alliance. In the end the issue was whether NATO would make its own defense decisions or whether the Soviets could enforce a major decision upon the alliance, and the alliance stuck to its guns. As a result of this cohesion, the United States was able to negotiate the INF Treaty with the Soviets.

But there was a price associated with the decision to proceed with the INF deployment, for it was necessary to agree to a reduction of 1,000 U.S. nuclear weapons from the stockpile deployed in Europe as part of the political trade-off for a favorable deployment decision. This trend continued with the 1983 Montebello decision to withdraw an additional 1,400 (mostly short-range) weapons in return for an agreement to pursue nuclear modernization. SHAPE's recent nuclear-weapon requirements study suggests still further reductions in the 1990s. Most of the weapons withdrawn prior to the INF Treaty were old and had limited military utility, for example, *Nike–Hercules* and atomic demolition munitions. But the P-IIs and GLCMs being withdrawn under the terms of the INF Treaty are modern, highly sophisticated systems. This represents a potentially significant loss in NATO's overall capability. And the new levels are very likely to constitute a de facto ceiling for the future.

Emerging Technologies

Another development that influences attitudes toward modernization of the stockpile of theater nuclear weapons today is the renewed interest in modernization of nonnuclear forces, particularly by the application of "advanced" or "emerging" technology. The report on *Discriminate Deterrence,* issued by the high-level, department of defense (DoD)-sponsored Commission on Integrated Long-Term Strategy, emphasized just such an approach to NATO planning.[13] The United States has been at the forefront of the effort to have NATO adopt new nonnuclear technologies and new doctrine, such as Deep Strike and Follow-On Forces Attack, to capitalize on these technologies. Predictably, the response from Europe has been mixed. On the one hand, the Europeans want to be responsive to U.S. initiatives, and they see some opportunities to gain new high-tech business for European industry. On the other hand, there is the traditional European suspicion that the United States is trying to decouple its nuclear deterrent from the defense of Europe with yet another initiative to substitute nonnuclear for nuclear forces in Europe's defense. Many Europeans, particularly Germans, see the deep-strike concept as provocative at a time when they want to strengthen political ties with the East. Some U.S. commentators, such as Steven Canby, have criticized the United States' reliance on technology to solve what are seen as essentially doctrinal and tactical problems.[14]

In any event, weapons with precision guidance are already being introduced into the inventories of both NATO and the Warsaw Pact. In the future, new sensors and communications systems will permit target acquisition at greater ranges and rapid and reliable communications between dispersed elements on the battlefield. But there is some doubt as to how rapidly these innovations will occur, particularly with defense budgets now declining. A major effort will be required to integrate new technologies and systems into the military establishments of both East and West. And it seems likely that the cost to develop and deploy advanced sensors, communications systems, munitions, and delivery platforms would have to be added to the already taxing 3 or 4 percent real growth rates in defense spending needed to achieve more modest conventional improvements. Furthermore, the prospect of replacing nuclear weapons with nonnuclear weapons in some important missions (e.g., against very hard or area targets) may not prove feasible for many years,[15] and even then nuclear weapons would still be required to deter nuclear use by the adversary or prevent nuclear intimidation or blackmail.

Some prominent American strategists have articulated a nuclear "no first use" (NFU) or "no early use" strategy.[16] Proponents of NFU contend that its adoption would not only strengthen public support for NATO policies and improve the chances for conventional force modernization, but would "reduce the risk of nuclear war . . . [and] improve the prospects for arms control." They recognize the difficulty of moving quickly to a dependable nonnuclear posture, although they strongly support such efforts as a means of restoring public confidence in NATO's defense program. The prospect of achieving conventional parity through CFE will increase the appeal of no first use, for many believe that parity will reduce the need to depend on nuclear weapons for deterrence and defense.

While the adoption of an NFU pledge could, over time, create a climate for negotiators to achieve more dramatic arms reductions, unless those negotiations are completely successful and virtually all nuclear weapons are eliminated—a continually elusive goal—an NFU pledge would not be meaningful. For by this group's own accounting, "so long as nuclear weapons remain available . . . the possibility of their use exists and cannot be discounted by the East or the West."[17] In addition, so long as NATO relies on the uncertainties created by nuclear weapons for deterrence, and on nuclear weapons to perform wartime missions not now achievable with nonnuclear forces, this proposal and others (e.g., to create a nuclear-free zone in Central Europe) could have a destabilizing effect in any transition phase. Indeed, this has been explicitly recognized by many Europeans, particularly in the FRG,[18] who feel their security would in fact be jeopardized by the implementation of such measures.

New Interest in Defenses

Yet another factor that has recently entered the picture is the renewed interest in defenses, particularly against ballistic missiles. The U.S. SDI program was initially a divisive factor in NATO because it was sprung on the allies abruptly without any advance consultations. This was followed by the rather curt offer from then Defense Secretary Weinberger for European participation, which most of his colleagues in Europe resented, in part because it left them with inadequate time for a response. So far as the Europeans were concerned, the program was off to a poor start.

Nevertheless, the debate proceeded and numerous trans-Atlantic discussions took place, and there is now a growing appreciation in Europe of several facts. First, there may be some interesting opportunities for European industry to participate in the SDI program. While initial expectations may well have been exaggerated on both sides of the Atlantic, European firms have acquired contracts and technical knowledge by participating in SDI. Few European governments have ignored these opportunities, and this is reflected in the formal agreements to participate in the program already made by our NATO allies in Britain, Germany, and Italy, as well as by the Japanese and Israeli governments.

Second, there is a growing acknowledgment in Europe of the large Soviet nuclear and nonnuclear missile threat that remains even after INF are removed.

While the direct applicability of the U.S. version of SDI to Europe is limited, the technologies developed in the SDI program may have applicability to both European air and missile defense. The potential of an antitactical ballistic missile (ATBM) at least needs to be explored. It is now clear that European defense experts favor a system, or systems, that deals with both the aircraft and missile threat, and which draws on U.S. technology.[19] Initial interest has focused on an improved version of the U.S. *Patriot* system.

Finally, there is a growing realization among Europeans that the Soviet Union and its allies are developing an ever-improving air defense system with ATBM capabilities (e.g., the SA-10 and SA-12) and that defense simply cannot be ignored in considering the European military balance.

Of course, there is also strong opposition to SDI and to ATBMs in Europe, as there is in the United States. This comes particularly from those who view defenses as disrupting the current arms control regime, complicating any future arms negotiations, and potentially leading to a new defensive arms race. These views are buttressed by an underlying skepticism about the feasibility of actually constructing an effective ballistic missile defense (BMD). Moreover, there are real concerns that a defensive system would lead to a new drain on resources—economic, technical, and manpower.

The Future of Flexible Response

A number of intermingling factors influence the roles and requirements for U.S. theater nuclear weapons in NATO. The interplay of these factors is a reflection of an environment in which the differences in perspective between the United States and its European allies cannot be fully resolved. The result is that the political health, indeed the continued existence, of the alliance depends on a compromise forged back in the mid-1960s that takes account of both American and European views. The result of this compromise is the strategy of flexible response, which has three elements: direct defense to meet the Pact attack directly; deliberate escalation, including escalating over the nuclear threshold; and general nuclear response, involving the use of U.S. strategic nuclear weapons.

Flexible response encompasses both the limited use of nuclear weapons and massive retaliation, and leaves unresolved exactly when and how escalation will occur. This has the virtue of keeping the enemy uncertain as to how NATO will act, thereby enhancing deterrence. However, it is a poor basis for concrete defense planning. It fails to provide a coherent rationale for employing NATO forces, and it could leave the allies in possibly fatal disarray should a war or crisis occur in Europe. Flexible response in its present form can justify almost any force posture, and it does. It leaves few clear guidelines as to how NATO would operate in time of war, and particularly the circumstances in which nuclear weapons would be employed. While there has been much study and planning of nuclear options by the NATO military staffs, little of this has been shared with or is well understood by the political leaders who would make the decisions to authorize the use of nuclear weapons.

Some very specific problems in implementing flexible response are exacerbated by the INF Treaty. The second element of flexible response, deliberate escalation, is particularly important to the European members of NATO. It could include the first use of theater nuclear weapons. The INF weapons contributed to deterring the Soviets by posing the threat of escalation and by the uncertainty they promoted with respect to how the West would respond to Soviet aggression. By removing the INF forces, the West removes its principal European-based capability to directly attack the Soviet homeland and takes away options on the escalation ladder. It also reduces Soviet uncertainties because it removes a key link between NATO theater forces and the U.S. strategic arsenal. The removal of INF also could affect the conventional battlefield—it removes a potential deep-strike threat and thus leaves the Soviets with less inhibitions on massing forces in rear areas.

The INF Treaty thus causes a fundamental dilemma. On the one hand, it increases the likelihood that NATO will have to resort to its nuclear arsenal. On the other, it takes away the heart of that arsenal and makes the transition to the second phase of flexible response all the more difficult. In recent testimony, SACEUR Galvin argued that flexible response can be accomplished, but its future success will be highly contingent on modernization of the remaining nuclear stockpile in Europe.[20] Such modernization seems increasingly problematical.

But flexible response, for all its lack of clarity—indeed because of the lack of clarity—has proven politically useful. It permits a consensus at a very general level that holds the alliance together, despite the differing priorities and aspirations of individual members. Efforts to improve upon it or to find a substitute have not proven feasible in the past. Thus, NATO proceeds with a strategy that is not wholly satisfying to anyone who has examined it in detail, but that seems preferable to any alternative yet devised. It seems unlikely, therefore, that radical changes will occur.

With these considerations in mind, we turn to an examination of the future requirements for theater nuclear weapons.

FUTURE REQUIREMENTS FOR THEATER NUCLEAR FORCES

Here, we examine the weapon characteristics, suggested future roles, and requirements. We first review the weapon requirements process and current service statements of requirements and the systems already identified to meet those needs. We then turn in the following section to a discussion of what we perceive as the factors most likely to constrain these future developments. By way of conclusion, we set forth three alternative future settings to examine how the interplay between requirements and constraints could shape in different ways the future development and deployment of nuclear weapons in Europe.

The Requirements Process

Ideally, the requirements for nuclear (or any other) weapons should be determined by a more or less logical process. Weapon requirements ultimately derive from strategy, for strategy defines military objectives and missions for a given threat. Objectives and missions, in turn, shape more precise operational concepts and requirements, which will yield appropriate system types and determine specific weapons.

In practice, however, the process is almost never as orderly or as structured as the ideal. Frequently, weapons requirements are established more because of technical feasibility than because military authorities had established a specific need. In some cases this "push" from the weapon developers has had beneficial results, producing weapons that turned out to fill a requirement. Many would argue that this was the case with the enhanced-radiation weapon, despite its disastrous political fate. In other cases, pushing a weapon through development with no clear military requirement resulted in unwanted weapons or weapons with unclear missions. Nuclear torpedoes are a prime example.

The timetable for the development of weapons also deserves a comment. Generally speaking, it tends to be governed by short-term considerations. While requirements for the next five years are fairly well understood and defined, those beyond five years are vague. Yet it is this future period that is of greatest interest because new designs typically involve long lead times—on the average it takes fifteen years for a tactical nuclear weapon to move from the concept development stage to field deployment.

In an effort to enhance longer-term R&D planning and to provide guidance to DOE and the national laboratories on potential nuclear-weapons needs, DoD revised in 1983 the Nuclear Weapons Development Guidance (NWDG). This process, and the document it produces, seeks to coordinate trends in technology with trends in national defense strategy and, particularly, nuclear-weapon employment policy. The newly instituted biennial NWDG process has improved the interaction between DoD officials concerned with requirements and DOE officials concerned with the development of weapons. Still, the resulting document is too short term in its outlook, lacks imaginative ideas, and often fails to capture future requirements adequately.[21]

Current Statements of Requirements

With respect to longer-range INF, the *Pershing* IIs and GLCMs deployed in Europe under the terms of the 1979 NATO dual-track decision are being dismantled in accordance with the terms of the 1987 INF Treaty. The medium-range P-IA of the United States has long ago been withdrawn, and the seventy-two P-IAs deployed by the West Germans will be withdrawn following the withdrawal of the missiles being removed under the terms of the INF Treaty. Decisions have been made by DoD recently to proceed with both a follow-on to the *Lance* (FOTL) system and a new tactical air-to-surface missile (TASM). However, NATO has yet to formally endorse deployment of these systems, which currently are scheduled to be operational in the mid-1990s. Without strong endorsement of these systems by NATO, the U.S. Congress is unlikely to continue supporting funding for procurement.

Current plans for short-range nuclear forces focus on replacement of a significant portion of the thirty-year-old W33 (8-inch) and twenty-three-year-old W48 (155 millimeters) nuclear artillery projectiles by the early 1990s. The W33 will be replaced by the W79, in production since 1981, and the W48 by the W82, now in the final stages of development. The Department of Defense notes that these new projectiles have greater accuracy, longer range (about 30 kilometers), higher reliability, and improved safety and security features. Consistent with congressional direction, all W82s, and those W79 rounds produced from 1984 on, will not be of the enhanced-radiation type.[22] In addition, Congress has placed a ceiling on the number of rounds to be produced: total W79 and W82 production may not exceed 925, well below SACEUR's stated requirement for modern AFAPs.

Improved safety and security is also a major focus of current nuclear modernization efforts. The army has programs in progress that include a secure weapon transport system to store and transport nuclear weapons with minimal exposure to terrorist and related threats; hardening of equipment against electromagnetic pulse; a secure storage concept of in-ground storage vaults and facilities to prevent or delay access to weapons; an improved upload system for more rapid dispersal;[23] and hardening of selected storage sites. Similar efforts are underway with respect to the munitions for NATO's nuclear-capable aircraft (e.g., the construction of storage vaults for nuclear weapons in the floors of hardened aircraft shelters).

Current efforts involving sea-based systems include the development of an improved gravity bomb for carrier-based aircraft; a new series of antisubmarine warfare weapons with better safety and security features for shipboard storage and with standoff capability; and a nuclear warhead—the W81—for the SM-2 surface-to-air fleet defense missile. While DoD projects that all of these weapons will be deployed by the "early 1990s," the Navy appears to be reluctant to field any new nonstrategic nuclear systems.[24] The navy is procuring the strategic sea-launched cruise missile (TLAM/N). Presently, however, there are no such systems committed to SACEUR, although various options are being considered.[25]

There have also been requirements mentioned (not necessarily for NATO) for a new atomic demolition munition, for a nuclear capability for the multiple-launcher rocket system, for multiburst weapons, for an advanced tactical air- or sea-launched cruise missile, for new naval nuclear weapons, and for other concepts. Finally, an effort is underway to find applications for an ATBM system, perhaps based on an improved *Patriot,* to counter the growing nonnuclear Soviet missile threat. It would be difficult to find support for deployment in today's political environment of a nuclear-armed theater missile defense. However, such a defense would be more effective in the near term than a nonnuclear system. More elaborate, multilayer systems capable of defending area targets would take longer to deploy, but could over time, either separately or in conjunction with a continental United States defense, take advantage of advanced technology developed in SDI.

It is clear that in the present environment few, if any, new nuclear systems will be developed for Europe. Some clear sense of priorities is therefore needed from the Bush administration. In this context, the following operational requirements or characteristics appear to be consistent with a strategy that continues to emphasize flexible response options, with a stress on the control of escalation and the discriminate use of nuclear weapons.

Improving the survivability of remaining theater nuclear forces, particularly the battlefield nuclear weapons and NATO's dual-capable aircraft, is a dominant requirement. Tactical nuclear-capable aircraft must be modernized to ensure their capacity to penetrate sophisticated Soviet air defenses. Stealth technologies will be important to ensure the survivability of these aircraft. Improved early warning is crucial, as is improved air defenses such as the *Patriot*.

Accuracy and discrimination will be of continued importance so long as escalation control remains part of NATO doctrine. This feature is also required so that the increasing number of hard targets (e.g., command centers) in the theater can continue to be held at risk effectively. Discriminate use implies a requirement for the lowest yield for accomplishing a particular targeting objective—indeed, this is a principal criterion in NATO target planning. Today, average warhead yields are too great and weapon accuracy not great enough for precision targeting.

Greater range is another desirable characteristic, particularly in short-range systems, for several reasons. First, it permits such systems to be deployed further back from the forward edge of the battle area. The new AFAP rounds contribute to this objective. Second, greater range is needed to provide support across corps and army boundaries. And third, it will help match the range of comparable Pact systems for effective counter-battery engagements.

All of the above requirements suggest lightweight, rugged warheads engineered with improved safety and reliability features so that they can be deployed quickly and widely in Europe with minimal risk of accident or seizure. Selectable-yield warheads will also be required to give NATO planners more flexible options.

In the future there will also be a growing need to cope with defenses—both air and missile defenses. This suggests more emphasis on several features: first, penetration aids for missiles, either of a physical or electronic variety. As defenses become more sophisticated, a terminally moving warhead will likely be required. The addition of penetration aids will likely increase the weight of a weapon. Thus, there will be a need to minimize warhead weight to the extent feasible to ensure adequate mobility of the weapon system. A second feature will be the employment of stealth technology where applicable. A third measure to cope with the increased threat of enemy defenses is to employ standoff air defense weapons, perhaps also utilizing stealth technology and possibly also nuclear arms.[26]

Constraints on Modernization

As suggested, there will be numerous constraints on theater nuclear weapon modernization because of budget constraints and a perception among many Europeans and Americans that the threat has diminished or is even disappearing. This view will manifest itself in a number of specific ways. In the near future there are likely to be constraints on resources dedicated to modernization, constraints on the peacetime movement of nuclear weapons, and constraints on European cooperation with respect to nuclear–weapon-related issues. Also, constraints on the sharing of nuclear technology will probably persist.

There will be *constraints on resources* for the modernization of theater nuclear weapons because of a general reluctance among many Europeans to devote limited resources to nuclear–weapon-related programs and a growing tendency by the U.S. Congress to condition support for new weapon programs on allied commitments to the program. European political leaders will be reluctant to take any initiatives to modernize these forces in the current political environment where the threat of war seems remote. Political leaders on the left are generally opposed to the modernization or enhancement of theater nuclear forces; indeed, they favor reduction or elimination of these weapons. Leaders on the right know it is an unpopular issue and prefer not to be seen as advocates. Resource constraints will interact with political constraints. Helmut Kohl, Margaret Thatcher, and others expended considerable political capital on the INF deployment issue and are in no hurry for another battle on that front.

Future leftist governments in the United Kingdom or the Federal Republic of Germany could press for removal of theater nuclear forces from their territories. Even short of this, there will be competing demands on resources that will constrain defense programs, and there will also be competing demands *within* defense budgets (i.e., nonnuclear modernization and, in the longer run, strategic defenses).

There will be *constraints on peacetime movement* and deployment of nuclear weapons due to the "nobody wants them here" syndrome. Also, the risk of terrorism will constrain peacetime movement. As we have discussed, this will make it more difficult to reduce the vulnerability of nuclear forces in Europe.

There will be *constraints on European cooperation* in nuclear arms. The French and British have made only limited progress at exchanging information. The question of German participation in nuclear matters is highly sensitive, and, at least at present, the Germans prefer that any nuclear link be with the United States rather than with the British or the French.[27] Present U.S.–FRG arrangements, of course, are under a dual-key system. Most Europeans probably prefer this too.

The United States also is *constrained in sharing nuclear technology* both by law and by policy. We are most unlikely to alter present dual-key arrangements. If European opposition to nuclear-weapon deployments increases, U.S. constraints could be tightened, and the threat of terrorism will make it more difficult to disperse weapons except in a deep crisis. U.S. transfer policies also could be affected by future arms control agreements. Finally, U.S. technology transfer policies today—which could change in the future—constrain transfer of important technologies related to nuclear weapons (e.g., high-speed computers).

Arms control limits could act as another constraint by further limiting the number or range of theater nuclear forces. Similarly, a new test ban could place additional restrictions on the development of weapons and the maintenance of the nuclear stockpile.

Given these constraints, it is most unlikely that all of the requirements outlined in the preceding section will be met. Yet it is very difficult to forecast just how these constraints will affect specific programs. To assist in illuminating the issues, we postulate below three possible alternative futures.

Alternative Political Settings

Future requirements for nuclear weapons will be driven by factors other than purely military considerations, and precisely gauging future trends is difficult, particularly when one looks beyond five years or so. A single projection would thus be misleading. Accordingly, in this section we discuss three possible alternative future settings to guide our thinking about future nuclear-weapon requirements. These scenarios encompass political, economic, and military factors.

The futures we postulate are not intended to encompass all possible alternatives. We recognize that there are other potential futures. Nonetheless, we have sought to choose futures that represent a spectrum of possibilities and highlight a wide range of variables likely to affect nuclear-weapon decision requirements through 2000.

Continuation of Current Trends

One course of events would be a direct extrapolation from current trends. The process of liberalization continues in Eastern Europe and the Soviet Union, but severe economic problems persist. The Soviet Union is able to contain nationalist pressures, but barely. Arms control negotiations progress under pressures to reduce defense spending in both East and West. Despite the relaxation of tensions, both NATO and the Warsaw Pact remain intact, although they preside over diminished force postures. A strategic arms reduction (START) agreement along the lines currently under discussion is reached. Strategic arsenals are cut by somewhat less than half, and both sides agree to a ten-year moratorium on space-based BMD deployments. Talks on conventional forces in Europe are given paramount priority by the U.S. administration and an initial CFE agreement is reached in 1991. Negotiations on SNF follow shortly and agreement is reached in 1994 to reduce SNF to a level of 2,000 weapons on each side. By 1995, START II and CFE II agreements are being negotiated. Defense budgets in the West declined by 2 percent in real terms.

about 100,000 troops from Europe and redirects some defense resources away from Europe and toward the Pacific region. Differences among the NATO members persist, particularly with respect to issues related to nuclear strategy, modernization, threat perception, and economic trade. Given only modest growth in the economies of most NATO countries, defense expenditures decline a point or two after inflation in the best of cases. The Europeans use these expenditures to modernize and improve their conventional forces, in part to compensate for the lost INF capability, but there is little support for the modernization of theater nuclear forces, and only limited resources are devoted to this effort.

Both superpowers continue to progress on their BMD programs, and the Soviets hint that, barring further progress in the post-START negotiations, they might expand their land-based BMD, currently limited to protection of the Moscow area. Both Britain and France warn that a Soviet BMD expansion will be met with increases in their independent strategic nuclear forces, but that is the extent of their reaction. Europe considers an ATBM program, but concludes that—beyond modest upgrading of *Patriot*—it is too costly. The allies adhere to flexible response as their doctrine, but they continue to fail in procuring the forces it requires. The West Europeans thus continue to "muddle through," generally perceiving the Soviet BMD threat as more of a concern to the United States than to themselves.

The Pact nations continue to suffer from slow economic growth and low levels of economic productivity, and the actual rate of growth of the Soviet defense budget begins to decline. Gorbachev's "peace offensive" aimed at the West continues in full gear and is greeted enthusiastically by the European allies. Throughout the 1990s, a number of trade agreements, loans, joint ventures, and cultural exchanges are reached between Eastern and Western Europe.

This scenario would see very little European support for modernization of U.S. nuclear forces in Europe, and it would lead to very limited modernization at best. Indeed, there is likely to be a reprise of the 1979 INF deployment decision, whereby further modernization is explicitly made contingent upon further reductions in the overall stockpile of short-range nuclear forces, and on the opening of arms control talks on these weapons. The 155-millimeter and 8-inch artillery shells may be gradually modernized, but at levels below SACEUR's stated requirements. Development proceeds on a corps-level FOTL, but deployment is deferred until after 2000. While development proceeds, a decision to deploy the Air Force's TASM continues to be delayed. Second generation stealth-capable cruise missiles are developed for both strategic and tactical missions, particularly for NATO's dual capable aircraft, but they are not expected to be deployed until early in the next century. The NATO stockpile shrinks to 2,000 weapons in 1998 as a result of the SNF agreement.

Reduced Tensions Erode NATO

In this scenario international tensions continue to subside although at a faster pace than in the scenario just described. The Soviet Union experiences little growth economically and suffers serious manpower shortages in the armed services. As a result, the Soviets cut back on military outlays in order to transfer resources to nonmilitary investment. The Soviets further accelerate their effort to negotiate arms control agreements that will permit major cutbacks in their military programs. As a result, there are major new treaties in which both sides agree to cut back their conventional forces in Europe by 50 percent below the CFE I levels; and to reduce their strategic offensive forces to 3,000 delivery vehicles by the year 2000. To facilitate this, the United States agrees to adhere to the narrow definition of the Antiballistic Missile (ABM) Treaty for ten years (i.e., until 2000). Within this narrow definition the Soviet Union proposes to negotiate possible deployments of limited numbers of land-based interceptors. The United States would like to deploy these systems to protect ballistic missiles and command and control facilities. The Soviets see utility in protecting their leadership. Both sides agree to strict limits on their antisatellite deployments.

In this environment there is no support for modernization of theater nuclear forces, in either the United States or in Europe. The Europeans do not perceive an imminent threat, and they cut back their defense budgets. United States defense budgets also are cut and 200,000 troops are removed from Europe. A Labour government in Britain has again reversed its position and is in the process of sharply curtailing the British nuclear program and eliminating all U.S. nuclear weapons from the United Kingdom. A Socialist government in the FRG has called for the elimination of all chemical weapons from Germany and the phasing out of nuclear weapons based on FRG territory by 1998. Leaders of the FRG indicate that they would like to meet with representatives of the GDR to discuss the establishment of a nuclear-free zone in the two Germanies. The United States expresses skepticism; the Soviets, joined by members of the GDR leadership, express their support for this initiative "in principle" and urge the immediate start of negotiations.

Despite the growth of antinuclear and antidefense sentiment in Europe, there has been little call for the formal dissolution of NATO. However, both NATO and the WTO have become skeletal organizations with no real substance. In response to the installation of leftist governments in Britain and West Germany, France assumes a greater role in alliance affairs and begins to work closely with the United States on defense-related matters. Nonetheless, calls for an ambiguously defined "Pacific orientation" mount from various groups within the United States, and such an orientation gains support in Congress.

Under these conditions the Europeans provide no support for the modernization of nuclear forces in Europe. With strong pressures to reduce nuclear forces in Europe and little support for compensatory improvements in nonnuclear forces, the United States agrees to the progressive elimination of theater-based nuclear forces in negotiations with the Soviet Union. It shifts its priorities to the development of a largely sea-based deterrent force. Both FOTL and TASM are halted by Congress when it becomes apparent that the allies will not support their deployment. Modernization efforts might therefore focus on advanced sea-launched cruise missiles and long-range air-delivered weapons for reaching targets in the Soviet Union. Weapons that can be quickly deployed from the continental United States are given high priority in the R&D process.

Increased Tensions Improve NATO Cohesion

This scenario envisages a dramatic increase in East–West tensions resulting from a change in leadership in the Soviet Union and the suppression of dissent within the Soviet Union. Talk of a Pacific orientation is still heard, but it is not taken seriously by the U.S. administration or by Congress. As a result, Europe remains the basis of U.S. planning. NATO becomes increasingly cohesive and takes various steps to address what is perceived to be a growing threat. Overall alliance defense spending is outpacing economic growth, although defense-spending increases remain unevenly distributed among the allies. In the United States, defense spending is particularly robust. Growth of the Soviet economy is barely evident, but Soviet defense spending nonetheless increases regularly. However, due to shortages of manpower and budget pressures, there is some reduction in the size of the Soviet armed forces to permit hardware improvements. Several Western nations, notably the Federal Republic of Germany, also face manpower constraints. Arms control negotiations are at a standstill, and both superpowers are pressing ahead rapidly to develop strategic defenses. The United States has an edge in technology, but little in the way of immediate deployable components. Intelligence indicates that the Soviets are on the verge of expanding their Moscow-based system and are preparing to give a six-month notice of their intention to withdraw from the ABM Treaty.

In this scenario the incentive to modernize obviously is high, and the increased levels of alliance defense spending makes it possible. However, most of the new defense spending is devoted to modernization of nonnuclear forces and strategic defense, the development and deployment of which is now given serious consideration among the Europeans. Theater nuclear forces get what is left over. To the extent that funds are spent for the modernization of theater nuclear forces, efforts are likely to focus on increasing the survivability of remaining theater nuclear forces and improving their penetrativity.

Modernized AFAPs are deployed in substantial numbers. A *Lance* follow-on is developed and deployed beginning in the mid-1990s, and TASM development would be accelerated. New, precision-guided bombs also are developed for tactical air. A ground-based ATBM is considered for Europe in the latter half of the decade.

Conclusions

Significant military requirements clearly exist for the modernization of the remaining theater nuclear forces in Europe. The current force falls far short of meeting the requirements of flexible response. In particular, the vulnerability of present theater nuclear forces makes it questionable that many would survive until the time that they might be needed. Nor would those that did survive necessarily be capable of controlled and discriminate use. This situation can become a potentially grave source of instability, and it is only exacerbated by the INF Treaty. Furthermore, the support systems for nuclear weapons as well as command and control procedures are in need of streamlining.

Yet it also seems clear that in the present political environment there is very little support for the modernization of theater nuclear forces in Europe, or even for measures such as greater peacetime dispersal and mobility that would make the current force less vulnerable. This is particularly troublesome as NATO increases its reliance on dual-capable aircraft, which operate from very vulnerable bases. While steps are being taken to reduce air base vulnerability and additional steps have been suggested, it does not seem likely that the situation will improve markedly in the near future.

On the other hand, the Soviets also face significant internal and alliance problems. Gorbachev recognizes the extent of the economic problems facing his country, and he seems genuinely committed to addressing them. He also seems prepared to support reform in Eastern Europe, even at considerable cost to the cohesion of the Warsaw Pact. He thus has strong incentives, based not on altruism but on fundamental self-interest, for wanting to reorient resources away from the military and toward the domestic sector. Indeed, doing so could strongly benefit the Soviet military in the long run, and some of the Soviet military leaders have acknowledged this. The Soviets can thus be expected to continue their long-standing efforts to erode both the cohesion of NATO and the West Europeans' political support for the U.S. presence in Europe. But it is doubtful that they would take any steps which could radically increase military tensions in the European theater.

Given the above, it will be crucial to think through a coherent rationale for theater nuclear forces that can help to shape priorities not only for the next generation of weapons, but for the systems which will follow those. It is unlikely that more than a few new systems will be approved for development and deployment over the next decade or so. It is possible that none could be. Cost considerations could take precedence over weapons design, particularly if the availability of special nuclear materials is constrained. It will be increasingly important, therefore, to develop weapons that can be used with more than one system. Likewise, it will be important to develop systems that can adapt to weapons perhaps not yet fully conceived. An example may be an advanced air- or sea-launched cruise missile with both tactical and strategic applications.

Yet, as suggested, desired development strategies will be constrained by a host of factors. The views of our allies will have increasingly greater weight on weapon-related decisions. We refer not only to the views of the West European leaders, but of the populace as well. The case for the modernization of theater nuclear forces, and indeed for these forces themselves, must be better made by American and European policymakers. They must be able to articulate a role for nuclear weapons if conventional parity is reached in Europe and the threat of short warning attack is sharply reduced. It must be conveyed that deterrence requires not only a sizeable nuclear arsenal but a flexible, militarily competent arsenal, capable of persuading an adversary that he would not be able to meet his wartime objectives, and, if he tried, he would be defeated. This rationale must be applicable to the new political conditions and the new military balance that is emerging in Europe. It also requires the will—and the ability to convey this will—on the part of all members of the Atlantic Alliance to support the modernization of these weapons and, in the worst of circumstances, to support the use of these weapons to protect their national interests.

It also seems likely that the U.S. Congress will continue to play an active role in weapon systems decisions. Along with the allies, their views also must be understood and taken into account. Finally, arms control will affect decisions on weapons. It is quite possible, for example, that by the turn of the century the United States and the Soviet Union will negotiate a new treaty that will sharply limit, or perhaps even eliminate, theater nuclear forces based in Europe. Also on the horizon is a comprehensive test ban or, barring that, a lowering of the test threshold or a limit on the number of tests each side is annually permitted. In sum, not only military requirements in the European theater, but political and economic trends will affect future weapons system research, development, acquisition, and deployment.

NOTES AND REFERENCES

1. In this chapter we refer to theater nuclear weapons and theater nuclear forces interchangeably. We also use standard Western definitions of terms such as *theater, tactical,* and *strategic,* recognizing that there could be some ambiguities in the precise meaning of these terms.
2. General Bernard Rogers (retired), testimony before the Committee on Armed Services, U.S. Senate, 100th Congress, 2nd sess., *NATO Defense and the INF Treaty,* pt. 2, February 1, 1988.
3. These recommendations include the development of a standoff air-to-ground missile, such as the Air Force's tactical air-to-surface missile, the development of a follow-on to the *Lance* missile, enhancement of NATO's dual-capable aircraft, and improvements in airbase survivability and in the dual-capable aircraft's ability to penetrate Soviet air defenses. He also recommended the modernization of NATO's 8-inch and 155-millimeter artillery. See Supreme Allied Commander John Galvin, testimony before the Committee on Appropriations, U.S. House of Representatives, 100th Congress, 2nd Sess., *Department of Defense Appropriations for 1989,* pt. 2, February 1, 1988.
4. For a useful history see David N. Schwartz, *NATO's Nuclear Dilemmas* (Washington, D.C.: Brookings Institution, 1983).
5. For example, McGeorge Bundy, George F. Kennan, Robert S. McNamara, and Gerard Smith, "Nuclear Weapons and the Atlantic Alliance," *Foreign Affairs* 60 (Spring 1982), pp. 753–68; and more recently, McGeorge Bundy et al., "Back from the Brink," *Atlantic* 258 (August 1986), pp. 35–41.
6. According to one source (using NATO data), the United States in 1985 "contributed nearly 75 per cent of the total $358 billion defense spending by alliance members. Britain contributed 6.2 per cent, France 5.4 per cent, and Germany 5.3 per cent. All the others together only add up to 7.7 per cent of the total." See Timothy W. Stanley, *Western and Eastern Economic Constraints on Defense: The Mutual Security Implications* (Washington, D.C.: The Atlantic Council, 1986).
7. See the comments by General Rogers in "NATO Chief Says Alliance Under Strain," *Washington Times,* September 11, 1986, p. 9. Also note Henry Kissinger's recent proposal that "some of the American forces not in Europe would contribute more effectively to global defense if they were redeployed as strategic reserves based in the United States and able to be moved to world trouble spots," in "Alliance Cure: Redeployment," *Washington Post,* May 13, 1986, p. A19.
8. See Tim Carrington, "Budget Woes Force U.S., Europe to Collaborate on New Weapons," *Wall Street Journal,* December 19, 1985, p. 32; David M. North, "NATO Members, Congress Boost Joint Weapons Development," *Aviation Week & Space Technology* (June 23, 1986), pp. 26–27.
9. David M. Abshire, "NATO on the Move," (Washington, D.C.: The Atlantic Council, 1985), *The Alliance Papers,* no. 6.
10. Quoted in Andrew J. Pierre et al., *Nuclear Weapons in Europe* (New York: New York University Press, 1984), pp. 98–100.
11. See Phillip A. Peterson and John G. Hines, "The Conventional Offensive in Soviet Theater Strategy," *Orbis* 27 (Fall 1983), pp. 695–739; Notra Trulock III, "Soviet Perspectives on Limited Nuclear Warfare," in Albert Wohlstetter, Fred Hoffman, and David Yost, eds., *Swords and Shields* (Lexington, Mass: Lexington Books, 1987).
12. For an elaboration, see C. N. Donnelly, "The Soviet Operational Maneuver Group: A New Challenge for NATO," *International Defense Review* 15 (1982), pp. 1177–86.
13. The Commission on Integrated Long-Term Strategy, *Discriminate Deterrence* (Washington, D.C.: Department of Defense, January 1988).
14. Steven L. Canby, "The Conventional Defense of Europe: The Operational Limits of Emerging Technology" (Washington, D.C.: The Wilson Center, 1984), working paper no. 55.

15. A group of RAND experts studying nonnuclear weapons concluded that "non-nuclear weapons can be conceived to pursue military objectives now assigned to nuclear weapons, but they are likely to be less easily developed or deployed for targets which do not have distinctive signatures, are mobile, hard, or very large in area or numbers. . . . The destructive effects of [nonnuclear weapons] are likely to be less certain, permanent, or observable than those of nuclear weapons; and they are likely to be much more susceptible to countermeasures." See Carl Builder et al., "The RAND Winter Study on Non-Nuclear Strategic Weapons: Executive Summary" (Santa Monica, California: RAND Corporation, 1984), N-2227-AF, p. 6.
16. Bundy et al., "Back from the Brink," pp. 35 and 40.
17. Ibid.
18. For example, Karl Kaiser, Georg Lever, Alois Mertes, and Franz-Joseph Shulze, "Nuclear Weapons and the Preservation of Peace," *Foreign Affairs* 60 (Summer 1982), pp. 1157–70.
19. For example, High Frontier Europa, "A First Report on European Defense Initiative," by the Von Hassell Policy Group, November 1985, (mimeographed); Stewart Menaul, "A European Defense Initiative," *Journal of Defense and Diplomacy* (February 1986), pp. 18–21.
20. Galvin said that "The strategy is one that is capable of accomplishment with the resources that we have available provided that, over time, those resources continue to be modernized . . . ." Hearings, Committee on Appropriations, House of Representatives, 100th Congress, 2nd Sess., *Department of Defense Appropriations for 1989*, pt. 2, February 1, 1988.
21. For a more detailed critical analysis of the NWDG, see Nuclear Concepts Working Group, "Evaluation of Long Term Planning for Nuclear Weapons Requirements," July 1986 (for Defense Nuclear Agency).
22. DoD, *Annual Report to the Congress FY 87*, p. 227.
23. Army Secretary John Marsh and Army Chief of Staff General John Wickham, "The Posture of the U.S. Army for FY 87," p. 34.
24. According to the former Assistant to the Secretary of Defense for Atomic Energy, Richard Wagner, "the Navy has not deployed a new non-strategic nuclear weapon since 1964." See David C. Morrison, "The Navy's Vanishing Nuclear Arsenal," *National Journal* (September 13, 1986), pp. 2184–85.
25. See statement by General Rogers in *Department of Defense Authorization for Appropriations for FY86 (Part 3: Unified Commands; Binary Chemicals)*, Hearings before Senate Armed Services Committee (Washington, D.C.: USGPO, 1985), p. 1452.
26. Continued development of standoff weapons will have to overcome an apparent reluctance on the part of some aviators to use these weapons. According to a recent press report, many "do not trust unmanned weapons and [they] want to preserve the tradition of foiling enemy defenses and dropping their bombs precisely." See *Washington Post*, September 26, 1986, p. A34.
27. However, in 1987, Jurgen Todenhoefer, a member of the *Bundestag* and a CDU–CSU spokesman on defense, called for the creation of an integrated European nuclear force including U.S., French, and British weapons and managed by an executive group consisting of the United States and all European members of NATO, including the FRG. "Somewhat to his surprise," according to an American who has followed these developments, "Todenhoefer received favorable reactions from a number of French and West German officials. [And], since the Spring of 1985, leaders of the Social Democratic Party have expressed a new interest in Franco–German nuclear cooperation." See Samuel F. Wells, "The United States and European Defense Cooperation," *Survival* (July–August 1985), p. 164.

Appendix

European Security in the Postwar Period

Documents from 1948–1989

The North Atlantic Treaty Organization

THE NORTH ATLANTIC TREATY[1]

The North Atlantic Treaty entered into force on August 24, 1949.

Preamble. The parties to this treaty reaffirm their faith in the purposes and principles of the UN Charter and their desire to live in peace with all peoples and all governments.

They are determined to safeguard the freedom, common heritage and civilization of their peoples, founded on the principles of democracy, individual liberty and the rule of law.

They seek to promote stability and well-being in the North Atlantic area.

They are resolved to unite their efforts for collective defence for the preservation of peace and security.

They therefore agree to this North Atlantic treaty.

Art. 1. The parties undertake, as set forth in the UN Charter, to settle any international dispute in which they may be involved by peaceful means in such a manner that international peace and security and justice are not endangered, and to refrain in their international relations from the threat or use of force in any manner inconsistent with the purposes of the United Nations.

Art. 2. The parties will contribute towards the further development of peaceful and friendly international relations by strengthening their free institutions, by bringing about a better understanding of the principles upon which these institutions are founded, and by promoting conditions of stability and well-being. They will seek to eliminate conflict in their international economic policies and will encourage economic collaboration between any or all of them.

Art. 3. In order more effectively to achieve the objectives of this treaty, the parties, separately and jointly, by means of continuous and effective self-help and mutual aid, will maintain and develop their individual and collective capacity to resist armed attack.

Art. 4. The parties will consult together whenever, in the opinion of any of them, the territorial integrity, political independence or security of any of the parties is threatened.

[1]Henry W. Degenhardt, *Treaties and Alliances of the World,* 3d ed. (London: Longman, 1981), pp. 166–67.

Art. 5. The parties agree that an armed attack against one or more of them in Europe or North America shall be considered an attack against them all, and consequently agree that, if such an armed attack occurs, each of them, in exercise of the right of individual or collective self-defence recognized by Art. 51 of the UN Charter, will assist the party or parties so attacked by taking forthwith, individually and in concert with the other parties, such action as it deems necessary, including the use of armed force, to restore and maintain the security of the North Atlantic area. Any such armed attack, and all measures taken as a result thereof, shall immediately be reported to the Security Council. Such measures shall be terminated when the Security Council has taken the measures necessary to restore and maintain international peace and security

Art. 6. For the purpose of Art. 5 an armed attack on one or more of the parties is deemed to include an armed attack on the territory of any of the parties in Europe or North America, on the Algerian Departments of France, on the occupation forces of any party in Europe, on the islands under the jurisdiction of any party in the North Atlantic area north of the Tropic of Cancer, or on the vessels or aircraft in this area of any of the parties.

[To provide for the accession to NATO of Greece and Turkey, Article 6 was reworded in a protocol to the treaty (signed on Oct. 22, 1951) as follows:

For the purpose of Art. 5 an armed attack on one or more of the parties is deemed to include an armed attack—(*i*) on the territory of any of the parties in Europe or North America, on the Algerian Departments of France, on the territory of Turkey, or on the islands under the jurisdiction of any of the parties in the North Atlantic area north of the Tropic of Cancer; (*ii*) on the forces, vessels, or aircraft of any of the parties, when in or over these territories or any other area in Europe in which occupation forces of any of the parties were stationed on the date when the treaty entered into force, or the Mediterranean Sea, or the North Atlantic area north of the Tropic of Cancer.

In January 1963 the clauses of this article referring to the Algerian Departments of France were recognized by the North Atlantic Council as having become inapplicable since July 3, 1962 (the date of Algerian independence).]

Art. 7. This treaty does not affect, and shall not be interpreted as affecting, in any way the rights and obligations under the Charter of the parties which are members of the United Nations, or the primary responsibility of the Security Council for the maintenance of international peace and security.

Art. 8. Each party declares that none of the international engagements now in force between it and any other of the parties or any third state is in conflict with the provisions of this treaty, and undertakes not to enter into any international engagement in conflict with this treaty.

Art. 9. The parties hereby establish a Council, on which each of them shall be represented, to consider matters concerning the implementation of this treaty. The Council shall be so organized so as to be able to meet promptly at any time. It shall set up such subsidiary bodies as may be necessary; in particular, it shall establish immediately a Defence Committee, which shall recommend measures for the implementation of Arts. 3 and 5.

Art. 10. The parties may, by unanimous agreement, invite any other European state in a position to further the principles of the treaty, and to contribute to the security of the North Atlantic area, to accede to this treaty. Any state so invited may become a party to the treaty by depositing its instrument of accession with the US Government.

Art. 11. This treaty shall be ratified and its provisions carried out by the parties in accordance with their respective constitutional processes. The instruments of ratification shall be deposited as soon as possible with the US Government, which will notify all the other signatories of each deposit. The treaty shall enter into force between the states which have ratified it as soon as the ratifications of the majority of the signatories, including those of Belgium, Canada, France, Luxembourg, the Netherlands, the United Kingdom and the United States, have been deposited, and shall come into effect with respect to other states on the date of deposit of their ratifications.

Art. 12. After the treaty has been in force for 10 years, or at any time thereafter, the parties shall, if any of them so requests, consult together for the purpose of reviewing the treaty, having regard for the factors then affecting peace and security in the North Atlantic area, including the development of universal as well as regional arrangements under the UN Charter for the maintenance of international peace and security.

Art. 13. After the treaty has been in force for 20 years, any party may cease to be a party one year after notice of denunciation has been given to the US Government, which will inform the governments of the other parties of the deposit of each notice of denunciation.

Art. 14. This treaty, of which the English and French texts are equally authentic, shall be deposited in the archives of the US Government. Duly certified copies will be transmitted by that government to the governments of the other signatories.

NORTH ATLANTIC COUNCIL COMMUNIQUÉ, DECEMBER 14, 1967: ANNEX TO COMMUNIQUÉ [HARMEL REPORT], FUTURE TASKS OF THE ALLIANCE[1]

Report of the Council

A year ago, on the initiative of the Foreign Minister of Belgium, the governments of the fifteen nations of the Alliance resolved to "study the future tasks which face the Alliance, and its procedures for fulfilling them in order to strengthen the Alliance as a factor for durable peace". The present report sets forth the general tenor and main principles emerging from this examination of the future tasks of the Alliance. . . .

3. The exercise has shown that the Alliance is a dynamic and vigorous organisation which is constantly adapting itself to changing conditions. It also has shown that its future tasks can be handled within the terms of the Treaty by building on the methods and procedures which have proved their value over many years.

4. Since the North Atlantic Treaty was signed in 1949 the international situation has changed significantly and the political tasks of the Alliance have assumed a new dimension. Amongst other developments, the Alliance has played a major part in stopping Communist expansion in Europe; the USSR has become one of the two world super powers but the Communist world is no longer monolithic; the Soviet doctrine of "peaceful co-existence" has changed the nature of the confrontation with the West but not the basic problems. Although the disparity between the power of the United States and that of the European states remains, Europe has recovered and is on its way towards unity. The process of decolonisation has transformed European relations with the rest of the world; at the same time, major problems have arisen in the relations between developed and developing countries.

5. The Atlantic Alliance has two main functions. Its first function is to maintain adequate military strength and political solidarity to deter aggression and other forms of pressure and to defend the territory of member countries if aggression should occur. Since its incep-

---

[1]United States Arms Control and Disarmament Agency, *Documents on Disarmament, 1967* (Washington, D.C.: U.S. Government Printing Office, 1967), pp. 679–81.

tion, the Alliance has successfully fulfilled this task. But the possibility of a crisis cannot be excluded as long as the central political issues in Europe, first and foremost the German question, remain unsolved. Moreover, the situation of instability and uncertainty still precludes a balanced reduction of military forces. Under these conditions, the Allies will maintain as necessary, a suitable military capability to assure the balance of forces, thereby creating a climate of stability, security and confidence.

In this climate the Alliance can carry out its second function, to pursue the search for progress towards a more stable relationship in which the underlying political issues can be solved. Military security and a policy of détente are not contradictory but complementary. Collective defence is a stabilizing factor in world politics. It is the necessary condition for effective policies directed towards a greater relaxation of tensions. The way to peace and stability in Europe rests in particular on the use of the Alliance constructively in the interest of détente. The participation of the USSR and the USA will be necessary to achieve a settlement of the political problems in Europe.

6. From the beginning the Atlantic Alliance has been a co-operative grouping of states sharing the same ideals and with a high degree of common interest. Their cohesion and solidarity provide an element of stability within the Atlantic area.

7. As sovereign states the Allies are not obliged to subordinate their policies to collective decision. The Alliance affords an effective forum and clearing house for the exchange of information and views; thus, each of the Allies can decide his policy in the light of close knowledge of each other's problems and objectives. To this end the practice of frank and timely consultations needs to be deepened and improved. Each Ally should play its full part in promoting an improvement in relations with the Soviet Union and the countries of Eastern Europe, bearing in mind that the pursuit of détente must not be allowed to split the Alliance. The chances of success will clearly be greatest if the Allies remain on parallel courses, especially in matters of close concern to them all; their actions will thus be all the more effective.

8. No peaceful order in Europe is possible without a major effort by all concerned. The evolution of Soviet and East European policies gives ground for hope that those governments may eventually come to recognise the advantages to them of collaborating in working towards a peaceful settlement. But no final and stable settlement in Europe is possible without a solution of the German question which lies at the heart of present tensions in Europe. Any such settlement must end the unnatural barriers between Eastern and Western Europe, which are most clearly and cruelly manifested in the division of Germany.

9. Accordingly the Allies are resolved to direct their energies to this purpose by realistic measures designed to further a détente in East–West relations. The relaxation of tensions is not the final goal but is part of a long-term process to promote better relations and to foster a European settlement. The ultimate political purpose of the Alliance is to achieve a just and lasting peaceful order in Europe accompanied by appropriate security guarantees.

10. Currently, the development of contacts between the countries of Western and Eastern Europe is now mainly on a bilateral basis. Certain subjects, of course, require by their very nature, a multilateral solution.

11. The problem of German reunification and its relationship to a European settlement has normally been dealt with in exchanges between the Soviet Union and the three Western powers having special responsibilities in this field. In the preparation of such exchanges the Federal Republic of Germany has regularly joined the three Western powers in order to reach a common position. The other Allies will continue to have their views considered in timely discussions among the Allies about Western policy on this subject, without in any way impairing the special responsibilities in question.

12. The Allies will examine and review suitable policies designed to achieve a just and stable order in Europe, to overcome the division of Germany and to foster European security. This will be part of a process of active and constant preparation for the time when fruitful discussions of these complex questions may be possible bilaterally or multilaterally between Eastern and Western nations.

13. The Allies are studying disarmament and practical arms control measures, including the possibility of balanced force reductions. These studies will be intensified. Their active pursuit reflects the will of the Allies to work for an effective détente with the East.

14. The Allies will examine with particular attention the defence problems of the exposed areas, e.g., the southeastern flank. In this respect the current situation in the Mediterranean presents special problems, bearing in mind that the current crisis in the Middle East falls within the responsibilities of the United Nations.

15. The North Atlantic Treaty area cannot be treated in isolation from the rest of the world. Crises and conflicts arising outside the area may impair its security either directly or by affecting the global balance. Allied countries contribute individually within the United Nations and other international organisations to the maintenance of international peace and security and to the solution of important international problems. In accordance with established usage the Allies or such of them as wish to do so will also continue to consult on such problems without commitment and as the case may demand. . . .

SPEECH OF M. LAURENT FABIUS, PRIME MINISTER OF FRANCE, AT THE INSTITUTE OF HIGHER NATIONAL DEFENCE STUDIES, SEPTEMBER 17, 1982 (EXCERPTS)[1]

. . . Ladies and Gentlemen, I have chosen, in my talk on our defence policy, to insist on the two main features that determine my Government's action:

- our defence is a factor of unity among the French,
- our defence is based on a permanent will for modernization.

Defence Policy, A Factor of Unity among the French

A very big majority of the French, whatever their political allegiance, accept the fundamental concept of nuclear deterrence.

In this respect, three principles seem to be particularly important: France's independence, solidarity with our allies, and the objective of curbing the excessive arms build-up.

France's Independence

As far as our territory and vital interests are concerned, we have to rely on ourselves. Our concept of deterrence in a relationship of unequal strength is both a token of France's credibility in the world and a factor contributing to the stability of the balance of force. This is why France spends much of her defence budget on acquiring an adequate nuclear capability. She can thus be sure, in all circumstances, of inflicting on a potential adversary enormous damage out of all proportion to what he might hope to gain from attacking her.

---

[1]*Survival* (November–December 1984), pp. 280–82.

This notion of nuclear adequacy implies that our arsenal be kept commensurate with that objective, which rules out following the USSR and US on a course of building up plethoric capabilities, a policy we cannot in any case afford. Nuclear adequacy implies that our capabilities must always remain credible, which means they must be protected, reliable and capable of penetrating evolving defence systems. It is this concept which has guided our options in the sphere of nuclear arms modernization, in particular the development of M4, medium-range air-to-ground and *Hades* missiles, or the decision to commission a seventh missile-launching nuclear submarine.

But our policy has a broader context. France's strategy is a non-war strategy. It rests on the idea that no nation can hope to win a conflict in Europe. Given the human concentration and the power of the opposing armaments, a major conflict, even if it were initially restricted to conventional arms, would in any event cause intolerable human and material losses.

In view of its irreversible and deliberately terrifying nature, this deterrent strategy cannot be diluted into any international alliance system. This is why the French deterrent can be brought into action only by the President of the French Republic, at the time and place he will have determined by reference to France's vital interests.

Solidarity

Independence without concessions and solidarity with our allies are complementary principles. The security of Europe, and therefore our own, depends on the cohesion of the Atlantic Alliance.

Regarding the Atlantic Alliance, . . .it is inappropriate to talk of France reintegrating NATO, for the simple reason that she has not left the Alliance, even though she left the integrated military command in 1966. The policy of gradually forging closer defence links with the Europeans does not call either the Alliance or our fundamental options into question. It consists in pragmatically strengthening the co-operation between Europeans on security matters, on a basis of concrete tasks. By doing so we are answering an expectation of European public opinions.

Outside the Atlantic Alliance, we also express our solidarity with other friendly countries who approach us to that effect, in different forms. Everywhere we seek one thing: to defend peace. . . .

To give even clearer expression to her capacity for solidarity with her allies, and friends, France now has, within the rapid action force, a number of professionalized units that can be deployed at very short notice. France does not have the means, and above all has no desire, to act as an international gendarme. She is simply intent on leaving her allies in no doubt as to her determination to honour her commitments wherever peace is seriously threatened.

Will to Reduce an Excessive Arms Build-up in the World

That will is not synonymous with weakness.

We wish to see constructive relations established with the USSR. Peace depends on this. But as the President of the Republic often points out, peace depends above all on a balance of force. That was why France took the view that the new "theatre" nuclear weapons deployed by the Soviets in Europe could not stay there without corresponding capabilities on the other side. Nevertheless, we consider it to be of the highest importance that a balance of force be established by negotiation at the lowest possible level. France therefore backs any attempt designed to achieve, through verifiable agreements, a reduction of the two super-powers' enormous nuclear potential.

We approve of any discussion between super-powers likely to rebuild confidence, but we do not accept that the terms of our own security be discussed in our absence. The President of the Republic stated before the United Nations General Assembly the three conditions set by France for participating in a negotiation between the countries possessing nuclear weapons:

- reduction of the difference in nature and quantity between the two great powers' arsenals and that of France;
- correction of imbalances in the sphere of conventional and chemical arms;
- no development of any new system that would destabilize the present foundations of deterrence and therefore of peace.

As regards the military use of space, France made realistic proposals last June aimed at getting negotiations under way to avoid a new spiral of the arms race. France wants "star peace". The projects currently under study in both the United States and the Soviet Union entail considerable risks, since each protagonist will be tempted to deploy additional offensive systems in response. There would be a risk of the development of anti-missile systems, both in space and on earth, being accompanied by a new upsurge in the offensive weapons sector. Lastly—and this, to us Europeans, is a decisive consideration—the development of offensive systems might trigger among our partners the feeling of an uncoupling of their security from that of the United States. The conditions would thus be created whereby those who enjoy conventional superiority in Europe might be tempted to exploit their advantage. The two super-powers, whose responsibility in this respect is considerable, must enter into negotiation on the subject. France's proposals may also enable the international community to take a hand in the matter, because it really does concern us all. . . .

Modernization of Our Defence

Ladies and Gentlemen, the second main feature of Government action, which is fully meaningful in defence policy, is the will to modernize France. A developed economy is either modern or useless. So is a defence capability. . . .

I should like you to think more especially about the issues that will dominate the years leading to the twenty-first century. We are not just living through a crisis, but through a world change. This fact, which underlies the Government's economic and social policy, must also underlie current military thinking.

For our military environment is changing radically. In the strategic context: achievement of strategic parity by the USSR and increase of its capabilities in Europe, resulting greater difficulty of maintaining the "coupling" between the United States and Europe; new vulnerabilities in the third world. In the political context: perception by European public opinion that decisions having life-or-death consequences for their countries can be taken elsewhere by anonymous machines. In the technological context: when the experts talk about "emergent technologies", they are referring to a veritable revolution in weapon systems likely in time to change strategic concepts.

The transformation that is bringing about a greater complexity of weapon systems is caused, as you know, by the fact that their environment is changing due to the introduction of very high performance micro-processing technology and to the development of the adversary's detection and targeting systems. From this, the experts are deducing possibilities of integrated management of the battlefield and of a much greater efficacy of non-nuclear weapons on the theatre of operations, or even in space. Hence the temptation to find replacements for nuclear deterrence, either by returning to the concept of battle on the "theatre", or by pursuing an illusory quest for 'defensive deterrence' based on veritable space shields that are supposed to be infallible against every form of aggression.

There is here, Ladies and Gentlemen, a challenge I ask you to think over. To abandon the bases of deterrence that have spared our world from world-wide conflict for forty years would be an act of deception. And advantage will certainly be taken of the understandable fear people have of nuclear weapons to present this deception to public opinions in the guise of technological progress. It would also be a mistake, of course, not to seek to integrate technological innovations in our defence capabilities. The real problem—and it is an essential subject of reflection with our allies—is that of placing technological transformations at the service of deterrence. Confronted with these developments, we must be firm about what is not to be tampered with and take care to discourage anything that might lead to a destabilization, but we must also be vigilant and have the technological means of developing the appropriate responses. In short, to preserve the credibility of our defence, we have to master modernity.

I have insisted on the modernity of our nuclear capabilities, but, clearly, our conventional ones are relevant. The discussion on the 1984–1985 Military Programme Act showed that the French wanted to provide for the development of our whole defence system. The Government has endeavoured to improve all aspects of military life. The land forces reorganization scheduled under the Act is designed to improve the efficacy of the combat forces. For instance, the Third Army Corps, numbering 50,000 men and 1,800 armoured vehicles including 400 battle tanks, has just established its command headquarters in Lille, while its tasks in a war have been reangled to cover the northern and north-eastern approaches to our territory. The measures already taken to improve the content of National Service have been well received by the French, as witness the interest shown in the "long service volunteer" scheme. This effort will be pursued, in particular by taking new measures to facilitate the return of national servicemen to civilian life at the end of their service. . . .

In the context of modernization, you will have to reflect, Ladies and Gentlemen, on practical means of gradually giving concrete substance to European co-operation. Some initiatives have already been taken. Whereas no major armaments programme was undertaken between 1970 and 1981, we have concluded several outline agreements (with Italy, Spain, Denmark, Greece, Belgium). It is through co-operation that our arms industries will acquire the human and financial resources with which to engage successfully in competition. We are developing this "variable geometry" co-operation with various partners. Some big projects are seeing the light of day: third-generation tactical missiles, multiple rocket launchers (MLRS), Franco–German combat helicopter for the nineties, European combat plane bringing five countries into association. . . . You also know that we are reinvigorating the joint thinking process in the convenient framework of Western European Union.

Our ties with Great Britain and Germany are particularly close. Our co-operation has entered the phase of implementation, whether as regards the combat helicopter, or a common military observation satellite project, or an in-depth dialogue on strategic concepts. I would even describe the Franco–German strategic dialogue as historic. Neither of the two partners is giving up the fundamental options of its security policy. But these leave room for a wide field of co-operation, be it on equipment—for instance the combat helicopter—or on the common military observation satellite project, or on the joint thinking exercise on concepts, for which, as you know, high-level Franco–German commissions have been set up. This policy is in fact part of a continuing process, since it implements the security clauses of the Elysée treaty concluded by Chancellor Adenauer and General de Gaulle. . . .

NORTH ATLANTIC COUNCIL COMMUNIQUÉ, DECEMBER 9, 1983 (EXCERPT)[1]

8. The successful conclusion of the Madrid Meeting contributes to the strengthening of the CSCE [Conference on Security and Cooperation in Europe] process. Although the Concluding Document agreed in Madrid falls somewhat short of the Allies' proposals, it is nonetheless substantive and balanced. The Allies attach equal importance to the implementation by all the CSCE countries of all provisions of both the Helsinki Final Act and the Madrid document, including their humanitarian aspects. An important result of the Madrid meeting was the agreement on a precise negotiating mandate for the Conference on Confidence and Security Building Measures and Disarmament in Europe (CDE) due to open in Stockholm in January 1984.

9. Since peace can and must be made more secure through equitable and verifiable agreements on disarmament and arms control, as well as through a balance of forces, the Allies have put forward a comprehensive series of proposals in this field. Their commitment to ensure security at the lowest possible level of forces was recently demonstrated by the member countries of the Nuclear Planning Group announcing at Montebello the withdrawal of 1,400 nuclear warheads from Europe, in addition to the reduction of 1,000 in 1980. The resulting stockpile will be the lowest in Europe for some twenty years. The Allies urge the Soviet Union to contribute to disarmament efforts in a concrete way and not to substitute declaratory proposals for deeds.

10. Ministers underline that the two-track decision of December 1979 by Allies concerned demonstrates the commitment of the Alliance to preserve peace and stability at the lowest possible level of forces. The Allies concerned reaffirm their commitment to pursue a balance of intermediate-range forces through arms control negotiations with the USSR. The progress report of the Special Consultative Group provides a comprehensive account of United States efforts, on the basis of close consultations with other involved Allies, to achieve such an agreement.

Unfortunately, the negotiations in Geneva have not yet achieved concrete results, due to the Soviet effort to maintain their monopoly of long-range land-based INF missiles. Deployments of Pershing II and ground-launched cruise missiles are thus proceeding in accordance with the December 1979 decision. Building on progress already achieved in Geneva, an agreement eliminating this entire category of United States and Soviet weapons, or at a minimum, limiting them to the lowest possible level, remains attainable. Allies concerned stress their willingness to continue negotiation and to halt, modify, or reverse the deployments now underway, on the basis of such an agreement. These Allies will continue to consult closely on steps directed toward achievement of this objective.

The Allies note with regret the unjustified Soviet decision to leave the negotiating table in Geneva. The United States has negotiated over the past two years while Soviet deployments of SS-20s have grown by over 300 warheads. The Allies believe that the Soviet Union shares an interest with the United States in reducing nuclear weapons through negotiated agreement. The Allies therefore call for bilateral negotiations on intermediate-range nuclear forces to resume as soon as possible.[2]

11. The Allies fully support the United States efforts in the Strategic Arms Reduction

---

[1]United States Arms Control and Disarmament Agency, *Documents on Disarmament, 1983* (Washington, D.C.: U.S. Government Printing Office, 1986), pp. 1031–33.

[2]Denmark and Greece reserve their positions on this paragraph. [Footnote in original.]

Talks (START), and they welcome the recent American initiatives which provide the basis for a significant build-down of US and Soviet strategic weapons. In particular, the United States has indicated its willingness to discuss trade-offs in areas of each side's particular interest and advantage in order to achieve an equitable agreement which promotes stability. The Allies call on the Soviet Union to respond to these initiatives in a positive spirit. In this connection they note with regret the recent Soviet failure to agree to a specific date for the opening of the next round of the talks.

12. The Conference on Confidence and Security Building Measures and Disarmament in Europe (CDE) is an important part of the CSCE process and provides new possibilities for increasing security throughout Europe. Allied countries will table a comprehensive package of concrete measures, in conformity with the mandate agreed in Madrid, designed to promote military openness in order to increase confidence and security and reduce the risk of surprise attack. They are resolved to negotiate actively for an early agreement on politically binding, militarily significant and verifiable measures which will cover the whole of Europe. As a sign of their determination, Ministers will themselves attend this Conference.

13. The Allies participating in the Mutual and Balanced Force Reductions (MBFR) talks reaffirm their determination to work for a mutually acceptable solution to the issues still barring progress. The Western draft Treaty presented last year provides a basis for a sound agreement. These Allies hope that the most recent Eastern statements indicate a new willingness to address the long standing basic issues of the Vienna negotiations and they are reviewing the state of these negotiations also in the light of these statements.

14. In the Committee on Disarmament, the Western participants continue to strive for concrete disarmament agreements. They consider as a priority task for this Committee the elaboration of a verifiable agreement banning the development, production and stock-piling of all chemical weapons.

The Allies remain gravely concerned about strong evidence of continued use of chemical weapons in South East Asia and Afghanistan, in violation of international law, and of Soviet involvement in the use of such weapons.[1] They welcome the fact that the United Nations is continuing to develop procedures to investigate allegations of the use of chemical weapons.

In the context of efforts aimed at the prevention of an arms race in outer space, the Allies have also proposed in the Committee on Disarmament that the existing international law concerning the peaceful use of outer space be reviewed.

LORD CARRINGTON, SECRETARY GENERAL OF NATO, ON THE ALLIANCE, SEPTEMBER 21, 1984 (EXCERPTS)[2]

. . . My main message would have borrowed from Pierre Harmel. . . . As he described it, the Alliance has two main functions: to maintain adequate military strength and political solidarity for deterrence and defence; and to pursue the search for progress towards a more stable relationship in which underlying political issues between East and West can be solved.

The formula is carefully drafted, and it is no accident that the word *political* appears twice. It is as true today as it was in 1967, both that the Alliance has a political as well as a military function; and that it will not succeed in its military function if it cannot maintain the

[1]Greece recalls its position as it has been expressed during the previous Ministerial sessions. [Footnote in original.]

[2]*Survival* (January–February 1985), pp. 38–41.

necessary political solidarity. So I see NATO as first and foremost a political organization; and the Alliance as one which must be politically as well as militarily strong if it is to do the job we expect of it.

. . .NATO, though undoubtedly a defensive alliance, is not an alliance concerned only with defence. Or, if it is, then it is with defence in a sense much wider than we normally use: because it embraces the political as well as the military aspects of what we do, both to keep the peace and to improve its quality. The politics are essential first of all to a proper assessment of what it is that we are trying to defend ourselves against; and secondly because security can be enhanced—or diminished—by political as well as by military means.

Assessing the Threat

The first of these activities—what in the jargon is called threat assessment—is crucial to what NATO does, and has been so since the Alliance was founded.

It is also at the heart of much of the political controversy which has enlivened our debates over the years. Indeed there are some who argue that it was a mistaken assessment of the threat which led to the signing of the North Atlantic Treaty in the first place; and others who claim, on the basis of their current assessment of Soviet policy, that individual policies of the Alliance, or the Alliance itself, are unnecessary or even dangerous in the world of today.

These are historical and political judgments. I could explain to you why I think them wrong, and I hope that I would be able to convince you. But I could not prove them to be wrong in the way that I can prove that H^2SO^4 is not the formula for drinking water.

What we call threat assessment, though it is based on as many facts as we can get and on sophisticated means of handling them, is not a science. We should beware of those who claim that it could be otherwise. More often than not, they will be crossing the important line between counting tanks as evidence of capability (which is a question of fact) and counting tanks as evidence of intention (which is a question of judgment).

We should not be afraid to acknowledge—and to make clear to an informed public—the extent to which the policy of the Alliance is based on judgment. At the same time, we should try to make it clear why we believe these judgments to be soundly based. Because, at the end of the day, what a democracy does to defend itself will be what the voting public accepts as a sensible precaution against perceived risk: a question of politics, if ever there was one.

The judgments we make are most unlikely to carry conviction if the member countries of the Alliance are seen to be helplessly divided on crucial parts of the assessment. Or if the diplomats and the military go off in their separate directions and concentrate their analysis on what may prove at best to be different parts of the same animal.

The problem is an important one for NATO. There is no doubt that major differences of assessment—and more particularly differences between the United States and Western Europe—could undermine the political solidarity which Harmel rightly identified as essential to the continuing military strength of the Alliance. It seems equally clear to me that the answer cannot be to brush points of difficulty under the carpet, in the hope that they will go away. They won't. What NATO must do is to bring together the information and insights of individual member countries: encourage the diplomats to cross-examine the military—and *vice versa*; keep up to date a body of agreed fact; and ensure that the judgments which emerge are ones which have survived a vigorous process of informed debate.

At one level, the Organization does this very well. There is at present a wide measure of agreement among member governments about what it is that we are facing. But if one looks beyond governments to Parliaments and public opinion more generally, the picture is less reassuring; and the Atlantic appears wider than it does at meetings of the NATO Council.

Part of the answer, as it seems to me, is to widen the debate within, as well as between, member countries. An essential ingredient of any successful policy towards the Soviet Union is a young man or woman with *Pravda* open on the desk and a volume of Russian history on the table at home. We need such people in the universities as well as in government; we need good lines of communication between them; and we should perhaps be more concerned than we are about what needs to be done to encourage the next generation.

East/West and West/West Relations

There are obvious limits to what NATO as such can do to help, and problems of resources and of access to classified information cannot be wished away. . . . Before I leave the question of assessment, let me highlight a point which is crucial to it: the importance of first hand contact with the Soviet Union at all levels, and with the Soviet leadership in particular. . . we do now need to take the measure of a Soviet leadership which is going through what—at the very least—must be a difficult period of adjustment after the deaths of Brezhnev and Andropov.

We have seen over the last few months a number of visits to Moscow by Western Foreign Ministers, and one by the President of France. These were not decision-making visits. Neither did they seek to convey an impression of greater warmth than exists at present in the different bilateral relationships in question. But they helped to fill the need to which I have pointed and I hope that such contacts will continue. I hope also that there will be travel in the other direction, and that more of the Soviet leadership will have occasion to see for themselves what we look like at home. . . .

To talk of West/West rather than transatlantic relations helps to underline the fact that the European members of the Alliance are not a monolith, and that differences between them can weaken the political solidarity of the Alliance even if they do not have a transatlantic aspect. But I have no doubt that it was the political solidarity which bridges the Atlantic which Harmel had most in mind when he wrote as he did. . . . The question of relations between Western Europe and North America is. . .crucially important to what we are trying to do as an Alliance; and I do not want to leave it without a more general word.

My generation is one which spent its political life in an atmosphere of transatlantic co-operation which had its origin in the war, and in the partnership which established the framework of the post-war world. I hope that we are not too sentimental about the bathwater, but we have grown rather attached to the baby. The strength which the democracies on both sides of the Atlantic draw from working together has not lessened with the passing of time. Neither has the importance of that partnership. I hope that politicians on both sides of the Atlantic will play their part in passing this message down to the next generation.

It is nowhere more important than when we come to look at East/West relations. There are, of course, differences of perception on either side of the Atlantic. They flow from differences of perspective which are in themselves quite natural. Western Europeans, for example, have a strong sense of living in a divided continent, and of what they have and had in common with the people to their East. For the United States, it is the Soviet Union—in a global rather than a purely European context—which presents the most clearly perceived image. My own conclusion is that the Western approach to East/West relations must take account of both these aspects of the truth if it is to be successful; that both the United States and Western Europe will be the losers if it is not successful; and that both must therefore make the effort to look at the picture in the round.

If that is done, the recipe for success can be simply stated: we need to define a sensible political strategy for East/West relations, and to stick to it. There is an English saying, that if we had some ham we could have some ham and eggs, if we had some eggs. I am not suggesting that my recipe will be an easy one to follow. But we can't afford not to try, and the picture is in fact rather less discouraging than some people would have us believe.

This is particularly true of the first half of the recipe, the definition of a sensible political strategy for East/West relations . . . an agreed political strategy is not so far away as some would have us think. Which is not to say that it can easily be put on paper. Like butterflies when you try to pin them down consensus tends either to escape or to end up with the colours blurred and the freshness lost. But let me sketch out five elements which I believe to be widely agreed by allied governments:

- The security which we seek for ourselves is not security at the expense of the Soviet Union. If the same is true of the Soviet Union, there is business to be done.
- Arms control and disarmament is not an alternative to security policy, but a part of it. We would like nothing better than to negotiate substantial reductions in the arms and armed forces which presently face each other in and across Europe. But we want to strike a balance which is fair and which we can count on to remain so. We will continue to do what is necessary to maintain the defences essential to our security.
- We have placed on the table specific proposals whose main points are worth recalling: a very substantial reduction in the strategic nuclear arsenals of the United States and the Soviet Union; the reduction of the intermediate range nuclear weapons to the lowest common ceiling that the Soviet Union will accept; a world-wide ban on chemical weapons; a common ceiling for armed forces in Central Europe at well below the current figures; and a series of practical measures to build confidence between the two sides. These proposals have not been put forward on a take it or leave it basis, and we are ready to negotiate. If they were all followed up, we believe that they could lead to an almost revolutionary improvement in the situation for both East and West.
- Important though it is, arms control and disarmament cannot bear the whole weight of East/West relations. Western governments, while remaining strongly critical of certain actions of the Soviet Union, are ready to play their part in a wider dialogue and to do so at the political level where appropriate; we favour the development of trade on prudent lines; we would like to see more individual contact between East and West; and we would like more generally to see progress in the implementation of the Helsinki Final Act. We do not reject the achievements of the 1970s: on the contrary, our aim is to build on what was soundly built a record of lasting achievement for the 1980s.
- There is a lot of work to be done and we should start now. We in the West are ready to do our share.

Sticking to a Strategy

But evolving a political strategy is only part of the job. There remains the second part of the recipe: the question of sticking to it. That, it seems to me, is where our difficulties mainly lie.

Before examining them in more detail, let me make it clear that by sticking to a strategy I do not mean sticking, whatever the circumstances, to every detail of a programme developed within it. . . . I think that it would have been quite wrong for the West to react to Afghanistan and Poland as if nothing untoward had happened. A strategy sound enough to deserve the name must make provision for an appropriate response to Soviet behaviour which flouts the rules on which a stable *modus vivendi* can be based; and the measures which Western countries took at that time fall within what I would call a sensible strategy.

Strategy can—and I would say must—be drawn wide enough to take account of different types of Soviet behaviour. But a strategy which is to be coherent cannot at the same time be drawn wide enough to accommodate major fluctuations in the Western approach to the underlying issues. The problem is one which has appeared in acute form at various times over the last ten years; and it will do so again if we cannot find the means to resolve it—or at least to manage it—rather better than at times we have.

Henry Kissinger, speaking in this context at the Palais d'Egmont earlier this year, called for a restoration of bipartisanship in the United States and for an end to what he called "the constant reassessments that disquiet our friends and confuse when they do not embolden our adversaries". He went on to say that the national interest does not change every four to eight years. I hope that those of you tempted to say "amen" to that will not regard the message as applicable to the United States alone. It will be meaningless to talk of consistency in political strategy if the Europeans are not prepared to be consistent; and that will require a good deal of heart-searching about bipartisanship on this side of the Atlantic too.

When I speak of bipartisanship, I do not in any way mean to exclude East/West relations—or defence policy, for that matter—from the agenda of political debate. But if it is our wish to exert some influence on East/West relations, rather than merely accommodating ourselves to what is brought about by the influence of others, then we must be prepared to develop policies which are good for the long haul; and to stay at the table for as long as it takes. The fact that Mr. Gromyko has been the Soviet Foreign Minister for 27 years is in a way exceptional; but, in a more important way, it illustrates a truth about the Soviet Union which we would be foolish to ignore.

It makes no historical or political sense to approach the Soviet Union in search of quick results. It makes even less sense to approach the Soviet leaders in search of a negotiating *quid pro quo* for something which they believe may fall into their laps for nothing next year, or the year after. How could anyone be expected to persuade the Politburo to make negotiating concessions to achieve equitable measures of nuclear disarmament, if Soviet Ambassadors were reporting that unilateral Western disarmament was just around the corner?

So what I have described as sticking to it will require both a wider consensus and one designed to take us into the future. Even if I am right in thinking that allied governments are now much closer than critics allow to a common appreciation of what we are up against, and to an agreed political strategy for East/West relations, that will not be enough. A political strategy more or less agreed today will not do us much good if it is thought likely to change at the next election—or even sooner, if it fails to deliver instant negotiating success against what I have suggested are improbable odds.

East/West relations is much more about where we want to be in ten years time than about where we want to be next Christmas. The political challenge is clear; and it is addressed to all of us in allied countries. Differences will no doubt continue to arise—both across the Atlantic and between government and opposition in individual member countries on matters of concern to the Alliance. We must find a way of keeping them to manageable proportions—and of keeping ourselves within the main lines of a coherent political strategy—if we are properly to fulfil the two functions defined by Harmel. And if we are to play our part in shaping the future of Europe. . . .

THE SECURITY POLICY OF THE FEDERAL GOVERNMENT [FEDERAL REPUBLIC OF GERMANY, 1985] (EXCERPTS)[1]

Atlantic Partnership and Strengthening of the European Pillar of the Alliance

. . .The consolidation and development of partnership between Europe and the North American allies is an essential goal of the policy of commitment to the Alliance adopted by the Federal Government. The North Atlantic Alliance and friendship with the United States of America are the foundations of German foreign and security policy. All of the allies share political fundamental convictions and are committed to civil and human rights. To protect these rights is the mission of the Western Alliance.

. . .The security partnership with the United States, a partnership which has developed over more than three decades, is vital for the Federal Republic of Germany. Without our alliance with the United States, security, peace, and freedom, are not conceivable for the Federal Republic of Germany. Therefore, close solidarity with the United States continues to be a prerequisite for any realistic German security policy. Close cooperation with the United States is the only way to lend weight and credibility to the German security policy vis-à-vis the Soviet Union.

. . .Without the military protection of the United States, West European security cannot be guaranteed. Even if the political conditions requisite to a pooling of the European forces were given, the political and military asymmetry of the European defence potential vis-à-vis the Soviet world power would continue to exist. The integration of North American and European defence capabilities is the only way of approximating parity with the military power of the Warsaw Pact. Atlantic security calls for mutuality. Much as Western Europe may have to rely on the protection of the United States, its independence and liberty and transatlantic partnership are of vital interest to the United States as well.

. . .Strengthening the European pillar in the Alliance is necessary to secure equilibrium between West and East. The Europeans will only be able to make this contribution if they enhance their cooperation. Essential elements of this cooperation are the Western European Union [WEU] and Franco–German cooperation. Both elements have been effectively promoted in recent years.

. . .The revitalization of the Western European Union has provided Western Europe with an effective instrument of enhancing cooperation in security matters within the North Atlantic Alliance. In the "Rome Declaration" of 27 October 1984 the foreign and defence ministers of the seven-member countries expressed their conviction that better utilization of WEU would not only contribute to the security of Western Europe but also to an improvement in the common defence of all the countries of the Atlantic Alliance. They reaffirmed their attachment to the goals of the modified Brussels WEU Treaty of 1954,

- to strengthen peace and security,
- to promote the unity and to encourage the progressive integration of Europe,
- to cooperate more closely both among member States and with other European organizations.

---

[1]The Federal Minister of Defence, *White Paper 1985: The Situation and the Development of the Federal Armed Forces* (Bonn, 1985), pp. 16–19, 21–24.

. . .The WEU Treaty has always been of special importance for the Federal Republic of Germany. The treaty obliges the contracting parties to render automatic and unrestricted aid and assistance in the event of an armed attack in Europe, leaves all military defence missions to NATO, and guarantees the stationing of the British Army on the Rhine in Germany. This guarantee and the resultant close cooperation with the United Kingdom in matters of defence policy form one of the bases of German security policy.

In the WEU Treaty the Federal Republic of Germany has undertaken to refrain from the production of nuclear, biological and chemical weapons, an undertaking which is a binding commitment under international law. As regards this undertaking, there has been no change. It was merely the German obligation to comply with certain restrictions on production of conventional defence commodities which was cancelled, and the quantitative controls of conventional weapons will be abolished by 1 January 1986.

. . .In the future, the main emphasis of WEU activities will be placed on substantive political discussions of the foreign and defence ministers who have decided to seek to harmonize their views on the specific conditions of security in Europe, in particular,

- defence questions,
- arms control and disarmament,
- the effects of developments in West–East relations on the security of Europe,
- Europe's contribution to the strengthening of the Atlantic Alliance, bearing in mind the importance of transatlantic relations,
- the development of European cooperation in the field of armaments in respect of which WEU can provide a political impetus.

WEU is also to consider the implications for Europe of crises in other regions of the world.

. . .The Federal Government regards the revitalization of WEU as an important step forward in European cooperation which has already proved its mettle in the EUROGROUP and IEPG [Independent European Programme Group]. This cooperation is to strengthen the European pillar within the Alliance, to open up a new important dimension in the process of European unification, and to contribute to the harmonization of views on any subject on which a common European position in the Alliance would be advisable. Those allies who are not members of WEU will be kept fully informed on WEU activities in the field of security policy.

A common European will is more necessary in these days than ever before. Only if this requirement is met will Western Europe continue to be a respected partner of the United States of America.

. . .Intensification of cooperation between the Federal Republic of Germany and France is of special political importance. It reflects political solidarity between nations that are neighbours and friends and are, together with all of the other European NATO countries, determined to improve European cooperation in matters of foreign and security policy.

The basis of Franco–German cooperation in security matters is the Elysée Treaty of 1963. The two governments have decided to intensify cooperation in security issues. With effect from October 1982, their foreign and defence ministers meet for summit talks twice a year.

Since that treaty has come into force, military cooperation, and in particular armaments cooperation, between the two countries has been considerably expanded. Their continual security dialogue impacts new impulses to Franco–German cooperation and European unification policy.

Arms Control and Disarmament

. . .Military strength alone is not enough to guarantee a peaceful future. Therefore, dialogue and cooperation in conjunction with arms control and disarmament on the basis of an assured deterrent and defence capability remain integral parts of our security policy within the Alliance. The Federal Government exploits any reasonable opportunity to attain security for all with less and less weapons, and to fulfill the hopes of mankind for a world stamped less by tension and being less highly armed.

. . .NATO's defensive strategy and its military posture which is exclusively tailored to ensuring our defence capability reflect appropriate self-restraint. The Federal Government has made an essential contribution to reducing NATO's nuclear capabilities unilaterally to the minimum of what is required to maintain credible deterrence. All the same, nobody can advocate unilateral measures of disarmament which would jeopardize the security of this country. It is only by patient and well-considered efforts aimed at the conclusion of mutually balanced verifiable agreements that we will be able to make any progress in arms control policy.

. . .A fundamental prerequisite for successful arms control efforts is readiness to enter into earnest negotiations. The Federal Government has regretted that the Soviet Union took advantage of the start of Western INF deployment in autumn 1983 to stop the nuclear arms control talks.

. . .This added considerably to the importance of the multilateral fora in which the Federal Government together with our allies continued the efforts at achieving more stability at lower force levels and with less military capabilities through arms control agreements:

- In January 1984, a new forum was established in Stockholm for the security policy dialogue between West and East with the inclusion of neutral and non-committed countries—the Conference on Confidence and Security-Building Measures in Europe. In close consultation the Western allies, in accordance with their mandate, made proposals for militarily relevant, politically binding, and adequately verifiable, measures whose area of application covers the whole of Europe from the Atlantic to the Urals. These measures are to diminish the risk of a surprise attack, reduce the possibility of political pressures by military force, and prevent miscalculations and misunderstandings. The aim of these measures is to assure the effectiveness of, and give expression to, the ban on the threat or use of force which is already established international law.
- In the Vienna Mutual and Balanced Force Reductions negotiations, the goal of the Federal Republic of Germany and her allies continues to be a balance of forces based on parity reached by means of common collective final ceilings for the personnel of the ground and air forces of both sides in Central Europe.[1] The essential aim is to utilize the Western negotiating proposal of July 1982, supplemented and completed on 19 April 1985, to solve inter alia the problems of the determination and verifiability of military strengths which continue to be a matter of dispute. The latest Warsaw Pact draft of a first partial agreement of 14 February 1984 concentrates on bilateral reductions in an initial phase and is hence aimed at achieving a limited intermediate result. The Federal Government is examining that proposal in close consultation with our

[1]The reduction area comprises the countries of Belgium, The Netherlands, Luxembourg, the Federal Republic of Germany, the GDR, Czechoslovakia, and Poland.

allies for positive elements which might be useful for a constructive continuation of negotiations.

- In the Geneva Committee on Disarmament (CD), the Federal Government continues to press in particular for progress in the efforts at achieving a worldwide, comprehensive, and reliably verifiable, ban on chemical weapons. In the eyes of the Federal Government, a ban on these weapons limited to Europe or parts of Europe would not be a way to improve our security situation and promote negotiations on a worldwide ban on chemical weapons. The Soviets have declared that they are ready in principle to consider the Western proposal of a permanent presence of international inspectors to witness the destruction of their stocks of chemical weapons. The announced detailed explanations of this proposal have, however, not yet come forth.

. . .The Federal Government expects the Soviet Union to demonstrate her declared willingness to recognize the legitimate security interests of others by acting accordingly in her new arms control negotiations with the United States.

The fact that intermediate-range weapons, strategic weapons, and space weapons will be discussed simultaneously in the Geneva negotiations is considered by the Federal Government as an important chance for the long-term stabilization of strategic relations. The Federal Government supports the goals of these negotiations which have been formulated by the two sides together: prevention of an arms race in space and termination of that race on earth, reductions of nuclear weapons, and consolidation of strategic stability.

The interest of the Federal Government is primarily focused on the limitation of intermediate-range nuclear weapons by appropriate arms control measures. With regard to these weapons it is essential to solve the problem with which the Soviet INF buildup confronts Western Europe and other regions adjacent to the Soviet Union by devising an arms control approach that provides for the elimination of the long-range Soviet and U.S. land-based intermediate-range missiles, or at least for their limitation as near to zero as possible.

. . .In view of the quantity and complexity of the problems to be solved, it is to be anticipated that a difficult and lengthy process of negotiation will ensue. The Federal Government considers it important that wherever there is a prospect of mutually acceptable solutions, agreements should be concluded even on partial problems.

The implications of bilateral arrangements between the United States and the Soviet Union for the security of the entire Alliance is the reason for our special interest in close consultations within the Western Alliance which will give all allies an opportunity to cooperate actively in shaping the character of the negotiations. The United States has agreed to such comprehensive consultations.

MAINTAINING SECURITY IN A CHANGING WORLD [UNITED KINGDOM, 1988] (EXCERPTS)[1]

The Dual Approach

. . .The last year has seen continued improvement in the atmosphere of East–West relations and important achievements for NATO's strategy for peace. At the end of 1987, the two

[1]*Statement on the Defence Estimates 1988,* vol. 1 (London: Her Majesty's Stationery Office, 1988), pp. 1–4.

elements of the Alliance's balanced security policy—strength in defence and readiness for dialogue—brought about the first-ever negotiated reduction in nuclear weapons. At their summit meeting in Brussels in March this year, NATO Heads of State and Government reaffirmed the validity of this dual approach, and reasserted the vital importance of the Alliance's strategy of deterrence as the guarantee of our collective security. It was thus a successful year for NATO and for British defence policy, confirming our policies on the course that we and our allies must follow to maintain security in a changing world.

. . .A year of unprecedented progress in arms control discussions between the United States and the Soviet Union culminated in the signature, in December 1987 at the Washington summit between President Reagan and Mr. Gorbachev, of the historic treaty eliminating US and Soviet ground-launched intermediate-range nuclear (INF) missiles. As the NATO summit agreed, this is a milestone in the Alliance's efforts to achieve a more secure peace, and the result of the political courage, realism and unity of its members. Soviet leaders were brought to the conference table because NATO deployed, in the face of much opposition, limited numbers of INF missile systems comparable to those already deployed by the Soviet Union. Unilateral gestures could never have achieved the same result. Indeed, had such gestures been made, we would face today the increased threat from Soviet missiles, with no countervailing force on the Western side.

The last year also saw the successful introduction of the confidence- and security-building measures agreed at the Stockholm Conference of 1986. And there was encouraging progress in other areas of arms control including, particularly, US–Soviet negotiations on strategic nuclear weapons, where there is now the prospect of agreement on a treaty reducing superpower armouries by 50%. . . .

Developments in the Soviet Union

. . .We are now at last beginning to see signs of change in the Soviet Union, reflected elsewhere in Eastern Europe. At home, Mr. Gorbachev has set in train a series of wide-ranging reforms. Abroad, he has argued the need for "new political thinking" in the nuclear and space age, and has laid stress on the growing interdependence of the countries of the world community. This "new thinking" acknowledges that there is more to security than military might and that Soviet security cannot be maintained on terms that inevitably mean insecurity for others.

. . .The British Government has welcomed the reforms that are taking place in the Soviet Union. More open internal policies, which bring the Soviet people more information about their own country and the world outside, are self-evidently desirable. And more pragmatic, less aggressive, external policies ought to make the Soviet Union a less uncomfortable neighbour. Certainly, as the Prime Minister's meeting with Mr. Gorbachev at RAF Brize Norton in December 1987 again showed, we are now able to pursue our bilateral dealings with the Soviet Union in an atmosphere that is more open and constructive than before.

. . .Although it would be imprudent to rely on a sustained change in Soviet attitudes, the prospect of its taking place is a challenge that the West must accept. The willingness of the present Soviet leadership to take part in the mutual concessions necessary to achieve an INF agreement is a good sign. But these are early days and we shall be looking for hard evidence, in the form of actions rather than words, that the "new thinking" represents a genuine and lasting shift in Soviet policy towards the West, and not merely a new approach to the presentation and execution of old policies. Progress in arms control has not yet been matched by comparable steps over issues such as human rights, equally important for the creation of lasting confidence between East and West. Soviet preparedness to withdraw from Afghanistan is a welcome step, but it is too early to tell whether this heralds a new Soviet approach to regional conflicts.

. . .Nor has there been any let-up in the Soviet Union's military modernisation programmes. . . . By the mid-1990s virtually the entire Soviet strategic nuclear force in place in the mid-1980s will have been replaced by new or modernised systems. And the Soviet Union is continuing to modernise its theatre nuclear forces in Europe, where even after the INF Treaty is implemented it will continue to enjoy a huge numerical advantage; improvements include the introduction of more accurate shorter-range nuclear missiles and the deployment of more advanced nuclear-capable aircraft, which will soon carry new nuclear-capable stand-off missiles. The Soviet Union continues to enjoy very substantial advantages in conventional forces, and is updating them so effectively that in key areas, such as aircraft and armour, the West's technological lead is being eroded. . . . Soviet capabilities for chemical warfare. . .far outstrip those of the West and cannot be justified by any comparable threat from NATO. These Soviet forces are deployed and structured in such a way as to give them a capability for offence and surprise.

. . .There is no reason to believe that Soviet leaders want war in Europe. But the Soviet Union has not, in the past, been reluctant to exploit the threat or use of force to achieve its political ends. And its avowed aim of removing all nuclear weapons from Europe would expose the countries of Western Europe to the intimidating power of Soviet superiority in conventional and chemical forces, which could be used in attempts to impose Soviet will. . . .

The NATO Response

. . .It was against this background that NATO leaders met in Brussels in March to take stock of the situation facing the Alliance and to set guidelines for future action. At that meeting they reaffirmed that:

- the link between the European and North American allies was vital to the security of each. . . .
- NATO defences should remain strong, recognising the crucial role of the nuclear deterrence provided not just by the United States' strategic deterrent but also by the presence of effective and up-to-date nuclear weapons in Europe. . . .
- they were willing to seek dialogue with the countries of the Warsaw Pact and further arms control agreements. . .based on their confidence in a sure defence.

In particular, all the NATO leaders agreed that, for the foreseeable future, there was no alternative to the Alliance's policy of deterrence based on an appropriate mix of adequate and effective nuclear and conventional forces, and expressed their resolve to keep them up to date where necessary.

. . .NATO's deterrent policy is carried out through the strategy of forward defence . . . and flexible response. Effective conventional forces will remain an important element in this strategy; but, however strong, they cannot achieve deterrence alone. So long as the West faces an adversary armed with nuclear weapons, it is self-evident that we must retain a similar capability to deter him from using or threatening to use them. More fundamentally, the need for the nuclear element in deterrence rests on the imperative to avoid any further war in Europe, conventional or otherwise; for even conventional weapons, of the power now available, would bring destruction on an unprecedented scale. The achievement of conventional parity would make an important contribution to enhancing international security. But even if an equal balance of conventional forces were to be reached, it could not be relied on to prevent aggression. History offers many examples of attacks launched against the odds; for example, when Hitler invaded the Soviet Union in 1941 his forces were substantially outnumbered. For the foreseeable future, only nuclear weapons will be able to confront a potential aggressor with risks so unacceptable that aggression is no longer a rational option.

. . .This reality is recognised in the arms control priorities endorsed at the NATO summit, and in particular in the reaffirmation by NATO leaders that further reductions in nuclear weapons, after the reduction in US and Soviet strategic arms now under negotiation, would come about only in conjunction with the establishment of conventional balance and the global elimination of chemical weapons.

The European Pillar

. . .A strong and stable partnership between the two sides of the Atlantic, resting on shared values and interests, lies at the heart of our security. As we explained in last year's Statement, neither North America nor Western Europe could carry the full burden alone. The presence in Europe of the conventional and nuclear forces of the United States plays an irreplaceable part in the defence of Europe and provides the essential linkage with the US strategic deterrent. Conversely, the credibility of allied defence cannot be maintained without a major European contribution.

. . .That European contribution is substantial. The defence expenditure of European NATO countries has steadily increased, and is now around one-third higher in real terms than it was in the early 1970s, enabling Europe to improve the quality of its forces. The European nations provide some 90% of the manpower, 85% of the tanks, 95% of the artillery, and 80% of the combat aircraft stationed in Europe, as well as 70% of the major warships readily available in the event of a conflict. European standing forces currently comprise over 3.5 million men and women, compared with US forces totalling 350,000 in Europe and 2.3 million world-wide. The active duty reserves of the European countries total 3.2 million men, compared with a US total of 1.7 million. Europe's aim is to make better use of the substantial resources it already devotes to the Alliance by more effective cooperation in defence and security matters; a more cohesive European effort can make a greater contribution to the Alliance than the sum total of our individual national efforts.

. . .There has been extensive activity in the field of European security cooperation during the last year. The potential of the revitalised Western European Union (WEU) was demonstrated by a series of meetings, under its auspices, which helped to concert contributions by member states to mine-clearance operations in the Gulf. . . . Although it has no role in either operational deployments or force planning, the WEU can provide a valuable forum for coordinating policy on such out-of-area issues.

. . .The WEU also met twice in 1987 at Ministerial level, in Luxembourg in April and in The Hague in October. These opportunities for both Defence and Foreign Ministers jointly to discuss European security issues are a unique feature of the WEU. The meeting in October was marked by the publication of the *Platform on European Security Interests,* which set out how the seven-member states intend to contribute to NATO's future security. It included their commitment to maintain credible and effective deterrence, based on both nuclear and conventional defences, and to pursue an active arms control policy aimed at enhancing security and fostering stability and cooperation in the whole of Europe. The platform will provide a basis for the development of the WEU as a forum for harmonising Western Europe's defence effort in support of the Alliance. The United Kingdom will continue to contribute constructively to this work when we take on the Presidency of the WEU in July this year. . . .

. . .The two meetings of **Eurogroup** Ministers in 1987 provided an opportunity for informal discussion of defence planning issues from a European perspective. . . .

. . .The European allies continue to strengthen their already close bilateral defence relations. At the Anglo–French summit in January 1988 a number of proposals for greater bilateral military cooperation were discussed, their purpose being to strengthen the collective defence arrangements of the Alliance. These include the further development of existing plans for wartime reinforcement that involve the use of French facilities, and the scope for closer French involvement in the forward defence of the Federal Republic of Germany. In addition, we have reached agreement with France on defence of the Channel fixed link and set up arrangements for obtaining better value for money in procurement. . . . The Government welcomes bilateral arrangements that contribute to the collective defence of NATO, and will continue to seek such opportunities wherever they arise.

DECLARATION OF THE HEADS OF STATE AND GOVERNMENT PARTICIPATING IN THE MEETING OF THE NORTH ATLANTIC COUNCIL IN BRUSSELS, March 2–3, 1988 (EXCERPTS)[1]

A Time for Reaffirmation

We, the representatives of the sixteen members of the North Atlantic Alliance, have come together to re-emphasize our unity, to assess the current state of East–West relations, to review the opportunities and challenges which lie ahead, and in doing so:

- to reaffirm the common ideals and purposes which are the foundation of our partnership;
- to rededicate ourselves to the principles and provisions of the Washington Treaty of 1949;
- to reassert the vital importance of the Alliance for our security, and the validity of our strategy for peace.

The Purposes and Principles of our Alliance

(2) Our Alliance is a voluntary association of free and democratic equals, united by common interests and values. It is unprecedented in its scope and success. Our security is indivisible. Our Alliance is dedicated to preserving peace in freedom and to collective self-defence, as recognised by the United Nations Charter. None of our weapons will ever be used except in response to attack.

(3) Our concept of a balanced security policy as set out in the Harmel Report has successfully stood the test of time. It remains valid in its two complementary and mutually reinforcing approaches: political solidarity and adequate military strength, and, on that basis, the search for constructive dialogue and co-operation, including arms control. The ultimate political purpose of our Alliance is to achieve a just and lasting peaceful order in Europe.

(4) The security in freedom and the prosperity of the European and North American

---

[1]*Survival* (May–June 1988), pp. 273–75.

Allies are inextricably linked. The long-standing commitment of the North American democracies to the preservation of peace and security in Europe is vital. The presence in Europe of the conventional and nuclear forces of the United States provides the essential linkage with the United States strategic deterrent, and, together with the forces of Canada, is a tangible expression of that commitment. This presence must and will be maintained.

Likewise, a free, independent and increasingly united Europe is vital to North America's security. The credibility of Allied defence cannot be maintained without a major European contribution. We therefore welcome recent efforts to reinforce the European pillar of the Alliance, intended to strengthen the transatlantic partnership and Alliance security as a whole.

The Atlantic Alliance cannot be strong if Europe is weak.

(5) Our aim will continue to be to prevent any kind of war or intimidation. By maintaining credible deterrence the Alliance has secured peace in Europe for nearly forty years. Conventional defences alone cannot ensure this; therefore, for the foreseeable future there is no alternative to the Alliance strategy for the prevention of war. This is a strategy of deterrence based upon an appropriate mix of adequate and effective nuclear and conventional forces which will continue to be kept up to date where necessary.

(6) While seeking security and stability at lower levels of armaments, we are determined to sustain the requisite efforts to ensure the continued viability, credibility and effectiveness of our conventional and nuclear forces, including the nuclear forces in Europe, which together provide the guarantee of our common security. Taking into account the structure of the Alliance, each of us undertakes to play his part in this joint endeavour in a spirit of solidarity, reaffirming our willingness to share fairly the risks, burdens and responsibilities as well as the benefits of our common efforts.

(7) We seek a just and stable condition of peace in which the sovereignty and territorial integrity of all states are respected and the rights of all individuals, including their right of political choice, are protected.

We want gradually to overcome the unnatural division of the European continent, which affects most directly the German people. We will continue to uphold the freedom and viability of Berlin and to support efforts to improve the situation there.

The search for improved and more stable relations with the Soviet Union and the other countries of Eastern Europe is among our principal concerns. We call upon these countries to work with us for a further relaxation of tensions, greater security at lower levels of arms, more extensive human contacts and increased access to information. We will continue the effort to expand co-operation with the East wherever and whenever this is of mutual benefit.

East–West Relations: The Way Ahead

(8) We have noted encouraging signs of change in the policies of the Soviet Union and some of its allies. . . .

(9) However, we have to date witnessed no relaxation of the military effort pursued for years by the Soviet Union. . . .

(10) . . .We are therefore determined to expand our practical co-operation in the field of armaments procurement and elsewhere. In this context we recognise the challenges to our industrially less advanced Allies and the need to address them through mutual assistance and co-operation.

(11) Arms control is an integral part of our security policy. We seek negotiations not for their own sake but to reach agreements which can significantly reduce the risk of conflict and make a genuine contribution to stability and peace. . . .

(13) The recently concluded INF agreement between the US and the Soviet Union is a milestone in our efforts to achieve a more secure peace and lower levels of arms. . . . We look forward to its early entry into force.

(14) Consistent with their security requirements, the fifteen Allies concerned will make use of all possibilities for effectively verifiable arms control agreements which lead to a stable and secure balance of forces at a lower level. For them, the comprehensive concept of arms control and disarmament includes:

- a 50% reduction in the strategic offensive nuclear weapons of the US and the Soviet Union to be achieved during current Geneva Negotiations;
- the global elimination of chemical weapons;
- the establishment of a stable and secure level of conventional forces, by the elimination of disparities, in the whole of Europe;
- in conjunction with the establishment of a conventional balance and the global elimination of chemical weapons, tangible and verifiable reductions of American and Soviet land-based nuclear missile systems of shorter range, leading to equal ceilings. . . .

(16) The resolution of East–West differences will require progress in many fields. Genuine peace in Europe cannot be established solely by arms control. It must be firmly based on full respect for fundamental human rights. . . .

(17) We agree that the speedy and complete withdrawal of Soviet troops from Afghanistan and the effective restoration of that country's sovereignty would be of major significance. . . .

(18) We hope that at their forthcoming summit in Moscow President Reagan and General Secretary Gorbachev will be able to build upon the progress achieved at the Washington meeting last December. We strongly support the efforts of the United States. . . .

(19) Reflecting upon almost four decades of common endeavour and sacrifice and upon the results achieved, we are confident that the principles and purposes of our Alliance remain valid today and for the future. We are united in our efforts. . . .

STATEMENT ON THE MINISTERIAL MEETING OF THE NORTH ATLANTIC COUNCIL IN MADRID ON JUNE 9–10, 1988[1]

1. At their meeting in Brussels on 2nd and 3rd March 1988, the Heads of State and Government of the Alliance reaffirmed its guiding principles and stressed their determination to continue working for the advancement of our common ideals and goals. It was in this spirit that at our meeting in Madrid we reviewed the international situation and the challenges and opportunities before us, taking into account recent positive developments.

2. Guided by our desire for a more peaceful and secure state of international affairs, we have continued since the Alliance Summit to consider the broad spectrum of issues concerning East–West relations and security, including arms control and the existing military force relationship. Against that background we discussed:

- the current situation in and prospects for Eastern Europe;
- the need for a substantial and balanced outcome of the CSCE follow-up meeting in

---

[1]*NATO Review* (June 1988), p. 30.

Vienna, at an early date, including significant progress on human rights and human contacts, and mandates for negotiations on conventional stability and confidence- and security-building measures;

- our continuing commitment to share equitably the risks, burdens and responsibilities, as well as the benefits of our common endeavour, and the need to renew our efforts to maintain, under evolving circumstances, a fair partnership, mindful of the structure of the Alliance.

3. We welcome the results of the Summit meeting in Moscow between President Reagan and General Secretary Gorbachev, both for their substance and as a portent for the future development of East–West relations. We welcome the unprecedented prominence accorded to human rights in the joint statement of the Summit, and hope that a more forthcoming Soviet attitude will also be reflected in the CSCE concluding document. We support the progress recorded towards an agreement on a 50% reduction in United States and Soviet strategic nuclear weapons, and stress the importance we attach to this objective in seeking security at lower levels of armaments. We welcome the entry into force of the INF Treaty as an important step in our search for effectively verifiable arms control agreements in accordance with the declarations of our Heads of State and Government.

4. The North Atlantic Council in Permanent Session has continued its consideration of the further development of the Alliance's comprehensive concept of arms control and disarmament as called for in the Statement issued in Reykjavik in June 1987. The Secretary General reported on the progress of this work and we look forward to receiving a written report at our next meeting in December.

5. We welcome the beginning of Soviet troop withdrawals from Afghanistan after over eight years of occupation. We hope that the Geneva Accords mark the start of a process which will enable the Afghan people to exercise their right to self-determination and enable their country to recover its full sovereignty and independence.

6. The maintenance of calm and stability in and around Berlin and the improvement of conditions there, as envisaged in the current Berlin initiative, remain key elements in East–West relations. As the EC's European City of Culture for 1988, Berlin is again demonstrating its vitality and attractiveness.

7. We greatly appreciate the hospitality of the Spanish Government and people on the occasion of our first meeting in Madrid. We take this opportunity to welcome again Spain's membership as yet another confirmation of the vitality of the North Atlantic Alliance. We also strongly support the process under way in response to proposals made by Spain for defining a significant Spanish military contribution to the common defence.

8. On completion of his term of office, we paid tribute to the departing Secretary General, Lord Carrington, for his outstanding contribution to the North Atlantic Treaty Organization. We expressed deep appreciation for his services to the strength and unity of the Alliance, and therefore to peace and freedom.

European Security Cooperation

TREATY[1] BETWEEN BELGIUM, FRANCE, LUXEMBOURG, THE NETHERLANDS AND THE UNITED KINGDOM OF GREAT BRITAIN AND NORTHERN IRELAND. SIGNED AT BRUSSELS, ON 17 MARCH 1948[2]

His Royal Highness the Prince Regent of Belgium, the President of the French Republic, President of the French Union, Her Royal Highness the Grand Duchess of Luxembourg, Her Majesty the Queen of the Netherlands and His Majesty the King of Great Britain, Ireland and the British Dominions beyond the Seas,

Resolved

To reaffirm their faith in fundamental human rights, in the dignity and worth of the human person and in the other ideals proclaimed in the Charter of the United Nations;

To fortify and preserve the principles of democracy, personal freedom and political liberty, the constitutional traditions and the rule of law, which are their common heritage;

To strengthen, with these aims in view, the economic, social and cultural ties by which they are already united;

To cooperate loyally and to coordinate their efforts to create in Western Europe a firm basis for European economic recovery;

To afford assistance to each other, in accordance with the Charter of the United Nations, in maintaining international peace and security and in resisting any policy of aggression;

To take such steps as may be held to be necessary in the event of a renewal by Germany of a policy of aggression;

To associate progressively in the pursuance of these aims other States inspired by the same ideals and animated by the like determination;

---

[1]Came into force on 25 August 1948 upon the deposit with the Belgian Government of the last instrument of ratification, in accordance with article X.

Following are the dates on which the instruments of ratification were deposited on behalf of each of the signatory States:

Belgium . . . 30 April 1948   Luxembourg . . . . . 10 June 1948
United Kingdom . 2 June 1948   Netherlands . 20 July 1948
France. . . . . 25 August 1948

[2]United Nations, *TS: Treaties and International Agreements Registered or Filed and Recorded with the Secretariat of the United Nations,* vol. 19, 1948, pp. 53–63.

Desiring for these purposes to conclude a treaty for collaboration in economic, social and cultural matters and for collective self-defence;

Have appointed as their Plenipotentiaries:

His Royal Highness the Prince Regent of Belgium,

His Excellency Mr. Paul-Henri Spaak, Prime Minister, Minister of Foreign Affairs, and His Excellency Mr. Gaston Eyskens, Minister of Finance,

The President of the French Republic, President of the French Union,

His Excellency M. Georges Bidault, Minister of Foreign Affairs, and His Excellency Mr. Jean de Hauteclocque, Ambassador Extraordinary and Plenipotentiary of the French Republic in Brussels,

Her Royal Highness the Grand Duchess of Luxemburg,

His Excellency M. Joseph Bech, Minister of Foreign Affairs, and His Excellency Mr. Robert Als, Envoy Extraordinary and Minister Plenipotentiary of Luxembourg in Brussels,

Her Majesty the Queen of the Netherlands,

His Excellency Baron C. G. W. H. van Boetzelaer van Oosterhout, Minister of Foreign Affairs, and His Excellency Baron Binnert Philip van Harinxma thoe Slooten, Ambassador Extraordinary and Plenipotentiary of the Netherlands in Brussels,

His Majesty the King of Great Britain, Ireland and the British Dominions beyond the Seas, for the United Kingdom of Great Britain and Northern Ireland,

The Right Honorable Ernest Bevin, Member of Parliament, Principal Secretary of State for Foreign Affairs, and His Excellency Sir George William Rendel, K.C.M.G. Ambassador Extraordinary and Plenipotentiary of His Britannic Majesty in Brussels, who, having exhibited their full powers found in good and due form, have agreed as follows:

Article I

Convinced of the close community of their interests and of the necessity of uniting in order to promote the economic recovery of Europe, the High Contracting Parties will so organize and coordinate their economic activities as to produce the best possible results, by the elimination of conflict in their economic policies, the coordination of production and the development of commercial exchanges.

The cooperation provided for in the preceding paragraph, which will be effected through the Consultative Council referred to in Article VII as well as through other bodies, shall not involve any duplication of, or prejudice to, the work of other economic organizations in which the High Contracting Parties are or may be represented but shall on the contrary assist the work of those organizations.

Article II

The High Contracting Parties will make every effort in common, both by direct consultation and in specialized agencies, to promote the attainment of a higher standard of living by their peoples and to develop on corresponding lines the social and other related services of their countries.

The High Contracting Parties will consult with the object of achieving the earliest possible application of recommendations of immediate practical interest, relating to social matters, adopted with their approval in the specialized agencies.

They will endeavour to conclude as soon as possible conventions with each other in the sphere of social security.

Article III

The High Contracting Parties will make every effort in common to lead their peoples towards a better understanding of the principles which form the basis of their common civilization and to promote cultural exchanges by conventions between themselves or by other means.

Article IV

If any of the High Contracting Parties should be the object of an armed attack in Europe, the other High Contracting Parties will, in accordance with the provisions of Article 51 of the Charter of the United Nations, afford the Party so attacked all the military and other aid and assistance in their power.

Article V

All measures taken as a result of the preceding Article shall be immediately reported to the Security Council. They shall be terminated as soon as the Security Council has taken the measures necessary to maintain or restore international peace and security.

The present Treaty does not prejudice in any way the obligations of the High Contracting Parties under the provisions of the Charter of the United Nations. It shall not be interpreted as affecting in any way the authority and responsibility of the Security Council under the Charter to take at any time such action as it deems necessary in order to maintain or restore international peace and security.

Article VI

The High Contracting Parties declare, each so far as he is concerned, that none of the international engagements now in force between him and any other of the High Contracting Parties or any third State is in conflict with the provisions of the present Treaty.

None of the High Contracting Parties will conclude any alliance or participate in any coalition directed against any other of the High Contracting Parties.

Article VII

For the purpose of consulting together on all the questions dealt with in the present Treaty, the High Contracting Parties will create a Consultative Council, which shall be so organized as to be able to exercise its functions continuously. The Council shall meet at such times as it shall deem fit.

At the request of any of the High Contracting Parties, the Council shall be immediately convened in order to permit the High Contracting Parties to consult with regard to any situation which may constitute a threat to peace, in whatever area this threat should arise; with regard to the attitude to be adopted and the steps to be taken in case of a renewal by Germany of an aggressive policy; or with regard to any situation constituting a danger to economic stability.

Article VIII

In pursuance of their determination to settle disputes only by peaceful means, the High Contracting Parties will apply to disputes between themselves the following provisions:

The High Contracting Parties will, while the present Treaty remains in force, settle all disputes falling within the scope of Article 36, paragraph 2, of the Statute of the International Court of Justice by referring them to the Court, subject only, in the case of each of them, to any reservation already made by that Party when accepting this clause for compulsory jurisdiction to the extent that that Party may maintain the reservation.

In addition, the High Contracting Parties will submit to conciliation all disputes outside the scope of Article 36, paragraph 2, of the Statute of the International Court of Justice.

In the case of a mixed dispute involving both questions for which conciliation is appropriate and other questions for which judicial settlement is appropriate, any Party to the dispute shall have the right to insist that the judicial settlement of the legal questions shall precede conciliation.

The preceding provisions of this Article in no way affect the application of relevant provisions or agreements prescribing some other method of pacific settlement.

Article IX

The High Contracting Parties may, by agreement, invite any other State to accede to the present Treaty on condition to be agreed between them and the State so invited.

Any State so invited may become a Party to the Treaty by depositing an instrument of accession with the Belgian Government.

The Belgian Government will inform each of the High Contracting Parties of the deposit of each instrument of accession.

Article X

The present Treaty shall be ratified and the instruments of ratification shall be deposited as soon as possible with the Belgian Government.

It shall enter into force on the date of the deposit of the last instrument of ratification and shall thereafter remain in force for fifty years.

After the expiry of the period of fifty years, each of the High Contracting Parties shall have the right to cease to be a party thereto provided that he shall have previously given one year's notice of denunciation to the Belgian Government.

The Belgian Government shall inform the Governments of the other High Contracting Parties of the deposit of each instrument of ratification and of signed the present Treaty and have each notice of denunciation.

IN WITNESS WHEREOF, the above mentioned Plenipotentiaries have affixed thereto their seals.

DONE at Brussels, this seventeenth day of March 1948, in English and French, each text being equally authentic, in a single copy which shall remain deposited in the archives of the Belgian Government and of which certified copies shall be transmitted by that Government to each of the other signatories.

For Belgium:
(*s*) P. H. SPAAK; G. EYSKENS

For France:
(*s*) BIDAULT; J. DE HAUTECLOCQUE
For Luxembourg:
(*s*) Jos. BECH; Robert ALS
For the Netherlands:
(*s*) B. V. BOETZELAER; VAN HARINXMA THOE SLOOTEN
For the United Kingdom of Great Britain and Northern Ireland:
(*s*) Ernest BEVIN; George RENDEL

PROTOCOLS TO THE 1948 BRUSSELS TREATY OF COLLABORATION AND COLLECTIVE SELF-DEFENCE AMONG WESTERN EUROPEAN STATES *SIGNED AT PARIS ON 23 OCTOBER 1954* *ENTERED INTO FORCE ON 6 MAY 1955* *DEPOSITARY: BELGIAN GOVERNMENT* (EXCERPTS)[1]

Protocol No. I Modifying and Completing the Above-Mentioned Treaty

His Majesty the King of the Belgians, the President of the French Republic, President of the French Union, Her Royal Highness the Grand Duchess of Luxembourg, Her Majesty the Queen of the Netherlands and Her Majesty the Queen of the United Kingdom of Great Britain and Northern Ireland and of Her other Realms and Territories, Head of the Commonwealth, Parties to the Treaty of Economic, Social and Cultural Collaboration and Collective Self-Defence, signed at Brussels on March the 17th, 1948, hereinafter referred to as the Treaty, on the one hand,

and the President of the Federal Republic of Germany and the President of the Italian Republic on the other hand, . . .

Have agreed as follows:

Article 1

The Federal Republic of Germany and the Italian Republic hereby accede to the Treaty as modified and completed by the present Protocol.

The High Contracting Parties to the present Protocol consider the Protocol on Forces of Western European Union (hereinafter referred to as Protocol No. II), the Protocol on the Control of Armaments and its Annexes (hereinafter referred to as Protocol No. III), and the Protocol on the Agency of Western European Union for the Control of Armaments (hereinafter referred to as Protocol No. IV) to be an integral part of the present Protocol.

Article 2

The sub-paragraph of the Preamble to the Treaty: "to take such steps as may be held necessary in the event of renewal by Germany of a policy of aggression" shall be modified to read: "to promote the unity and to encourage the progressive integration of Europe".

. . .

---

[1]Jozef Goldblat, *Agreements for Arms Control: A Critical Survey* (London: Taylor and Francis, 1982), pp. 145–48.

Article 3

The following new Article shall be inserted in the Treaty as Article IV: "In the execution of the Treaty the High Contracting Parties and any organs established by Them under the Treaty shall work in close co-operation with the North Atlantic Treaty Organization".

. . .

Article 6

The present Protocol and the other Protocols listed in Article I above shall be ratified and the instruments of ratification shall be deposited as soon as possible with the Belgian Government.

They shall enter into force when all instruments of ratification of the present Protocol have been deposited with the Belgian Government and the instrument of accession of the Federal Republic of Germany to the North Atlantic Treaty has been deposited with the Government of the United States of America.

. . .

Protocol No. II on Forces of Western European Union

. . .

Article 1

1. The land and air forces which each of the High Contracting Parties to the present Protocol shall place under the Supreme Allied Commander Europe in peacetime on the mainland of Europe shall not exceed in total strength and number of formations:

(*a*) for Belgium, France, the Federal Republic of Germany, Italy and the Netherlands, the maxima laid down for peacetime in the Special Agreement annexed to the Treaty on the Establishment of a European Defence Community signed at Paris, on 27th May, 1952;[1] and

(*b*) for the United Kingdom, four divisions and the Second Tactical Air Force;

(*c*) for Luxembourg, one regimental combat team.

2. The number of formations mentioned in paragraph 1 may be brought up to date and adapted as necessary to make them suitable for the North Atlantic Treaty Organization, provided that the equivalent fighting capacity and total strengths are not exceeded.

3. The statement of these maxima does not commit any of the High Contracting Parties to build up or maintain forces at these levels, but maintains their right to do so if required.

Article 2

As regards naval forces, the contribution to NATO Commands of each of the High Contracting Parties to the present Protocol shall be determined each year in the course of the Annual Review (which takes into account the recommendations of the NATO military authorities). The naval forces of the Federal Republic of Germany shall consist of the vessels and formations necessary for the defensive missions assigned to it by the North Atlantic

---

[1]This Treaty, signed by Belgium, France, FR Germany, Italy, Luxembourg and the Netherlands, did not enter into force.

Treaty Organization within the limits laid down in the Special Agreement mentioned in Article 1, or equivalent fighting capacity.

. . .

Article 5

The strength and armaments of the internal defence and police forces on the mainland of Europe of the High Contracting Parties to the present Protocol shall be fixed by agreements within the Organization of Western European Union, having regard to their proper functions and needs and to their existing levels.

. . .

Protocol No. III (with Annexes) on the Control of Armaments

Part I. *Armaments not to be manufactured*

Article 1

The High Contracting Parties, members of Western European Union, take note of and record their agreement with the Declaration of the Chancellor of the Federal Republic of Germany (made in London on 3rd October, 1954, and annexed hereto as Annex I) in which the Federal Republic of Germany undertook not to manufacture in its territory atomic, biological and chemical weapons. The types of armaments referred to in this Article are defined in Annex II. These armaments shall be more closely defined and the definitions brought up to date by the Council of Western European Union.

Article 2

The High Contracting Parties, members of Western European Union, also take note of and record their agreement with the undertaking given by the Chancellor of the Federal Republic of Germany in the same Declaration that certain further types of armaments will not be manufactured in the territory of the Federal Republic of Germany, except that if in accordance with the needs of the armed forces a recommendation for an amendment to, or cancellation of, the content of the list of these armaments is made by the competent Supreme Commander of the North Atlantic Treaty Organization, and if the Government of the Federal Republic of Germany submits a request accordingly, such an amendment or cancellation may be made by a resolution of the Council of Western European Union passed by a two-thirds majority. The types of armaments referred to in this Article are listed in Annex III.

Part II. *Armaments to be controlled*

Article 3

When the development of atomic, biological and chemical weapons in the territory on the mainland of Europe of the High Contracting Parties who have not given up the right to produce them has passed the experimental stage and effective production of them has started there, the level of stocks that the High Contracting Parties concerned will be allowed to hold on the mainland of Europe shall be decided by a majority vote of the Council of Western European Union.

Article 4

Without prejudice to the foregoing Articles, the types of armaments listed in Annex IV will be controlled to the extent and in the manner laid down in Protocol No. IV.

Article 5

The Council of Western European Union may vary the list in Annex IV by unanimous decision.

. . .

Annex I

The Federal Chancellor declares:

that the Federal Republic undertakes not to manufacture in its territory any atomic weapons, chemical weapons or biological weapons, as detailed in paragraphs I, II and III of the attached list,

that it undertakes further not to manufacture in its territory such weapons as those detailed in paragraphs IV, V and VI of the attached list. Any amendment to or cancellation of the substance of paragraphs IV, V and VI can, on the request of the Federal Republic, be carried out by a resolution of the Brussels Council of Ministers by a two-thirds majority, if in accordance with the needs of the armed forces a request is made by the competent Supreme Commander of the North Atlantic Treaty Organization;

that the Federal Republic agrees to supervision by the competent authority of the Brussels Treaty Organization to ensure that these undertakings are observed.

Annex II

This list comprises the weapons defined in paragraphs I to III and the factories earmarked solely for their production. All apparatus, parts, equipment, installations, substances and organisms, which are used for civilian purposes or for scientific, medical and industrial research in the fields of pure and applied science shall be excluded from this definition.

I. Atomic Weapons

(*a*) An atomic weapon is defined as any weapon which contains, or is designed to contain or utilise, nuclear fuel or radioactive isotopes and which, by explosion or other uncontrolled nuclear transformation of the nuclear fuel, or by radioactivity of the nuclear fuel or radioactive isotopes, is capable of mass destruction, mass injury or mass poisoning.

(*b*) Furthermore, any part, device, assembly or material especially designed for, or primarily useful in, any weapon as set forth under paragraph (*a*), shall be deemed to be an atomic weapon.

(*c*) Nuclear fuel as used in the preceding definition includes plutonium, Uranium 223 [*sic*], Uranium 235 (including Uranium 235 contained in Uranium enriched to over 2.1 per cent by weight of Uranium 235) and any other material capable of releasing substantial quantities of atomic energy through nuclear fission or fusion or other nuclear reaction of the material. The foregoing materials shall be considered to be nuclear fuel regardless of the chemical or physical form in which they exist.

II. Chemical Weapons

(*a*) A chemical weapon is defined as any equipment or apparatus expressly designed to use, for military purposes, the asphyxiating, toxic, irritant, paralysant, growth-regulating, anti-lubricating or catalysing properties of any chemical substance.

(*b*) Subject to the provisions of paragraph (*c*), chemical substances, having such properties and capable of being used in the equipment of apparatus referred to in paragraph (*a*), shall be deemed to be included in this definition.

(*c*) Such apparatus and such quantities of the chemical substances as are referred to in paragraphs (*a*) and (*b*) which do not exceed peaceful civilian requirements shall be deemed to be excluded from this definition.

III. Biological Weapons

(*a*) A biological weapon is defined as any equipment or apparatus expressly designed to use, for military purposes, harmful insects or other living or dead organisms, or their toxic products.

(*b*) Subject to the provisions of paragraph (*c*), insects, organisms and their toxic products of such nature and in such amounts as to make them capable of being used in the equipment or apparatus referred to in (*a*) shall be deemed to be included in this definition.

(*c*) Such equipment or apparatus and such quantities of the insects, organisms and their toxic products as are referred to in paragraphs (*a*) and (*b*) which do not exceed peaceful civilian requirements shall be deemed to be excluded from the definition of biological weapons.

Annex III

This list comprises the weapons defined in paragraphs IV to VI and the factories earmarked solely for their production. All apparatus, parts, equipment, installations, substances and organisms, which are used for civilian purposes or for scientific, medical and industrial research in the fields of pure and applied science shall be excluded from this definition.

IV. Long-range Missiles, Guided Missiles and Influence Mines

(*a*) Subject to the provisions of paragraph (*d*), long-range missiles and guided missiles are defined as missiles such that the speed or direction of motion can be influenced after the instant of launching by a device or mechanism inside or outside the missile, including V-type weapons developed in the recent war and subsequent modifications thereof. Combustion is considered as a mechanism which may influence the speed.

(*b*) Subject to the provisions of paragraph (*d*), influence mines are defined as naval mines which can be exploded automatically by influences which emanate solely from external sources, including influence mines developed in the recent war and subsequent modifications thereof.

(*c*) Parts, devices or assemblies specially designed for use in or with the weapons referred to in paragraphs (*a*) and (*b*) shall be deemed to be included in this definition.

(*d*) Proximity fuses, and short-range guided missiles for anti-aircraft defence with the following maximum characteristics are regarded as excluded from this definition:
Length, 2 metres;
Diameter, 30 centimetres;
Speed, 660 metres per second;
Ground range, 32 kilometres;
Weight of war-head, 22.5 kilogrammes.

V. Warships, with the exception of smaller ships for defence purposes

Warships, with the exception of smaller ships for defence purposes are:

(*a*) Warships of more than 3,000 tons displacement;

(*b*) Submarines of more than 350 tons displacement;

(*c*) All warships which are driven by means other than steam, Diesel or petrol engines or by gas turbines or by jet engines.

Annex IV

VI. Bomber aircraft for strategic purposes

List of Types of Armaments to be Controlled

1. (*a*) Atomic
   (*b*) biological, and
   (*c*) chemical weapons.

In accordance with definitions to be approved by the Council of Western European Union is indicated in Article I of the present Protocol.

2. All guns, howitzers and mortars of any types and of any rôles of more than 90 mm. calibre including the following component for these weapons, viz., the elevating mass.

3. All guided missiles.

Definition: Guided missiles are such that the speed or direction or motion can be influenced after the instant of launching by a device or mechanism inside or outside the missile; these include V-type weapons developed in the recent war and modifications thereto. Combustion is considered as a mechanism which may influence the speed.

4. Other self-propelled missiles of a weight exceeding 15 kilogrammes in working order.

5. Mines of all types except anti-tank and anti-personnel mines.

6. Tanks, including the following component parts for these tanks, viz:
(*a*) the elevating mass;
(*b*) turret castings and/or plate assembly.

7. Other armoured fighting vehicles of an overall weight of more than 10 metric tons.

8. (*a*) Warships over 1,500 tons displacement;
   (*b*) submarines;
   (*c*) all warships powered by means other than steam, diesel or petrol engines or gas turbines;
   (*d*) small craft capable of a speed of over 30 knots, equipped with offensive armament.

9. Aircraft bombs of more than 1,000 kilogrammes.

10. Ammunition for the weapons described in paragraph 2 above.

11. (*a*) Complete military aircraft other than:
    (*i*) all training aircraft except operational types used for training purposes;

(*ii*) military transport and communication aircraft;

(*iii*) helicopters;

(*b*) air frames, specifically and exclusively designed for military aircraft except those at (*i*), (*ii*) and (*iii*) above;

(*c*) jet engines, turbo-propeller engines and rocket motors, when these are the principal motive power.

Protocol No. IV on the Agency of Western European Union for the Control of Armaments

Part I. *Constitution*

Article 1

The Agency for the Control of Armaments (hereinafter referred to as "the Agency") shall be responsible to the Council of Western European Union (hereinafter referred to as "the Council"). It shall consist of a Director assisted by a Deputy Director, and supported by a staff drawn equitably from nationals of the High Contracting Parties, Members of Western European Union.

. . .

Part II. *Functions*

Article 7

1. The tasks of the Agency shall be:

(*a*) to satisfy itself that the undertakings set out in Protocol No. III not to manufacture certain types of armaments mentioned in Annexes II and III to that Protocol are being observed;

(*b*) to control, in accordance with Part III of the present Protocol, the level of stocks of armaments of the types mentioned in Annex IV to Protocol No. III held by each member of Western European Union on the mainland of Europe. This control shall extend to production and imports to the extent required to make the control of stocks effective.

2. For the purposes mentioned in paragraph 1 of this Article, the Agency shall:

(*a*) scrutinise statistical and budgetary information supplied by members of Western European Union and by the NATO authorities;

(*b*) undertake on the mainland of Europe test checks, visits and inspections at production plants, depots and forces (other than depots or forces under NATO authority);

(*c*) report to the Council.

. . .

Part III. *Levels of stocks of armaments*

Article 13

1. Each member of Western European Union shall, in respect of its forces under NATO authority stationed on the mainland of Europe, furnish annually to the Agency statements of:

(*a*) the total quantities of armaments of the types mentioned in Annex IV to Protocol No. III required in relation to its forces;

(*b*) the quantities of such armaments currently held at the beginning of the control years;
(*c*) the programmes for attaining the total quantities mentioned in (*a*) by:
(*i*) manufacture in its own territory;
(*ii*) purchase from another country;
(*iii*) end-item aid from another country.

2. Such statements shall also be furnished by each member of Western European Union in respect of its internal defence and police forces and its other forces under national control stationed on the mainland of Europe including a statement of stocks held there for its forces stationed overseas.

3. The statements shall be correlated with the relevant submissions to the North Atlantic Treaty Organization.

. . .

Parties: Belgium, France, FR Germany, Italy, Luxembourg, Netherlands, UK

FRANCO–GERMAN SUMMIT STATEMENT, February 27–28, 1986 (EXCERPTS)[1]

. . .(1) The President of the French Republic and the Chancellor of the Federal Republic made a wide-ranging survey of the security and defence problems which concern France and the Federal Republic. They declared their agreement to promote all measures which can contribute, in the framework of the Treaty of the Elysee of 22 January 1963, to strengthen the already close ties which bind the two countries in this field.

(2) Operational Co-operation, Co-operation between the Armed Forces

The governments of the two countries have agreed to increased operational co-operation between their armies.

With a view to increasing efficiency in strategy and inter-operability, the President of the French Republic and the Chancellor of the Federal Republic authorize the pursuit of studies relating to the improved use of French forces in Germany, and in particular the Force d'Action Rapide. From 1986–7 appropriate joint manoeuvres will be held to realize this agreement.

(3) Training

The President of the French Republic and the Chancellor of the Federal Republic have furthermore agreed to projects for the joint training of officers. . . . It is envisaged that from 1986 officers will join together for common training courses conducted alternately in France and Germany.

(4) Consultation

Within the limits imposed by the extreme rapidity of such decision the President of the French Republic is prepared to consult the Chancellor of the Federal Republic of Germany on

[1]*Survival* (July–August 1986), pp. 366–67.

the possible use of French pre-strategic weapons on German territory. He reminds us that in this matter, the decision cannot be shared.

The President of the Republic announces that he has decided with the Chancellor of the Federal Republic of Germany to set up the technical means for immediate and reliable consultation in times of crisis. . . .

Declaration Concerning the *Hermes* project

The German government agrees to examine with the French government the conditions and extent of its participation in the *Hermes* project. The final decision of the German government will be taken after further bilateral consultations in Autumn 1986, so that the realization of the project can begin in mid-1987, all conditions being fulfilled, as was decided at the European Space Agency Conference held in Rome in January 1985.

The two governments examined the progress made in the development of the Franco–German combat helicopter. They welcome the favourable report of the armament directors.

The two governments are favourable to a renewal and widening of the *Airbus* range.

Realizing the strategic and financial importance of this choice, they agreed to ask their manufacturers, in conjunction with their other partners, to deepen their studies of the choice of machine to develop, calendar of research and development and production and of the industrial and financial framework.

Once proposals have been made by the manufacturers, the ministers responsible will meet promptly.

SPEECH BY SIR GEOFFREY HOWE, BRITISH FOREIGN SECRETARY, MARCH 16, 1987 (EXCERPTS)[1]

To my generation, who remember the Second World War, the need for defence is self-evident. Nor is there much dispute about the source or the nature of the threat. We recognise that 40 years of post-war peace and security in Europe have been made possible by the Atlantic Alliance—by a strategy solidly based on deterrence, and since 1967, as a result of the report of a distinguished former Belgian Foreign Minister, Pierre Harmel, combined with the promotion of confidence, contacts, and dialogue between East and West. Last but not least, those who remember the war can readily accept that US nuclear protection, and the presence of US troops in Europe, is at the heart of our security strategy.

But for the two generations on both sides of the Atlantic who have grown up with no recollection of Europe at war, the threat is less obvious. In the United States, Europe no longer dominates military thinking. And at the strategic level, the SDI research programme is extending the frontiers of debate.

Thus, the old consensus has become less clear-cut. The increasing divergence of views, not least among the political parties in my own country, brings a need to justify again, and sometimes to reconsider, the basic assumptions which we have long taken for granted. As we consider how to prolong forty years of European peace into the next century, that may be no bad thing. . . .

Europe no longer dominates American thinking as much as it did in the past. The Pacific

---

[1]*Survival* (July–August 1987), pp. 376–80.

theatre, South-west Asia and Central America now take a larger share of the United States' time and attention than they did. There are calls in Congress and outside for more emphasis on worldwide maritime and airlift capabilities.

There are other pressures in Washington as well. In his recent speech at Davos my German colleague Hans-Dietrich Genscher commented that there is sometimes an impression given by the NATO Alliance that 240 million Americans are defending more than 300 million Europeans against 280 million Russians. The sizeable European defence effort which I have described is often not fully appreciated in the United States. We Europeans must do better at explaining what we do contribute. But burden-sharing pressures are unlikely to disappear, not least because they are exacerbated by the US deficit, and by an increasingly protectionist trade atmosphere. This makes it all the more important that trade disputes should not be allowed to spill over into the strategic relationship.

A distaste for reliance on nuclear weapons is not a new phenomenon in America. It is not in the American nature to be happy when held hostage to an irresistible threat. Some have long questioned whether the US would ever be prepared to use nuclear weapons in response to a Warsaw Pact conventional attack in Europe. Geographically, not being faced with an obvious conventional threat, many Americans see clear attractions in strategic defences.

I do not mean to suggest that the United States Administration, and indeed Americans more widely, do not remain fully committed to NATO strategy. It does, and they do. But we need to be alert to trends in American thinking which might diminish our security—perhaps not today or tomorrow, but possibly in the longer term.

Meanwhile there have also been changes in Europe. In the UK and the FRG, the two major members of the integrated military structure, there is no longer a common view between the political parties on the role of nuclear weapons in defence policy. From the point of view of NATO this is regrettable, although since the great majority of people in the UK still support the need for a British nuclear deterrent, it is not likely to do my political opponents anything but harm in electoral terms. Certainly, the British Government continues to see a need for nuclear weapons for the foreseeable future.

Apart from the nuclear debate, we have also recently heard calls for purely "defensive defence". I find it hard to understand this philosophy. Are the tens of thousands of Warsaw Pact tanks a sign of purely defensive intentions? Does European public opinion really want to abandon forward defence? Would we prefer the battle to be fought in our own centres of population? And is it right that an aggressor should be immune from risk of retaliation against his own country? Surely not.

But even when we agree on the need for defence we still have to find the resources. The very success of NATO in keeping the peace for 40 years makes it ever harder as time goes by to secure adequate resources from electorates who have never known war. The costs of new weapons systems continue to grow rapidly. The Soviet Union has so far shown no sign of reducing the 13–15% of GDP which it devotes to defence. On the contrary, its defence spending continues to grow at some 3% per annum in real terms. Even there I have some hope that the Soviet Union may at last be feeling mounting economic pressure, and that this may help to explain Mr Gorbachev's seemingly more urgent approach to arms control.

The shifts in attitudes in Europe and in the US are not likely to lead to immediate and dramatic changes in Alliance strategy. But I believe that it is right (and prudent) to prepare for the possibility, in the longer term, of a "less nuclear" world. That is to say, a continuation of a strategy of nuclear deterrence but one that is based on smaller quantities and a narrower range of systems. And I believe that we should recognise too a greater responsibility on the part of Europeans for the defence of Western Europe—in other words for a more truly equal second pillar of the Alliance.

Looking to the future I should like to identify a number of issues which need to be addressed, so that we manage the future security of the Alliance in a planned way rather than find unwelcome developments thrust upon us.

For the foreseeable future Europe's security will depend on nuclear weapons. For our agreed policy of Flexible Response to be credible, an adequate nuclear capability is required, of a certain variety and mix. But there is no magic in particular numbers or systems. We do not envisage fighting a battlefield or theatre nuclear war confined to Europe. And our concept of deterrence is not based on a pre-determined progression through specific layers of nuclear escalation. As things stand, if we can reduce numbers of weapons on the basis of a satisfactory quid pro quo from the Warsaw Pact we should certainly do so. We must, of course, keep a careful eye on the resulting balance; but there is clearly much scope for negotiated reductions in the present excessive levels of nuclear weapons.

That is why I believe that an INF treaty between the United States and the Soviet Union, provided that its detailed provisions on issues like verification and shorter-range systems are watertight, will be an important security gain. It will show that firmness of political purpose and Allied solidarity pay off. It will mean the elimination of a whole class of nuclear weapons which on the Soviet side posed a major and unique threat to Western Europe. Even with the removal of the cruise missiles and the *Pershing* IIs from NATO's armoury, we shall still have more than enough nuclear weapons to maintain that element of uncertainty about our reactions which is crucial to deterrence. To talk of the denuclearization of Western Europe when there are still some 4,000 or so nuclear warheads deployed here is absurd. And to cast doubt on the political solidarity, or coupling, of the American commitment to the defence of Europe when there are still over 325,000 troops on our soil is unnecessarily to turn fancy into fear. On this latter point I endorse what the Supreme Allied Commander Europe, General Rogers, said when he addressed this Institute two weeks ago.

But we need to look beyond the prospective INF agreement, to identify our aims for the shorter-range missiles in Europe. We need to determine whether strategic stability would be enhanced or undermined by a greater reliance on air breathing systems for certain types of mission. . . .

On conventional weapons, we need to be clear which Warsaw Pact armaments constitute the most serious risk to our security and to shape our negotiating position accordingly. . . .

But common defence means common responsibilities. Britain, the FRG, the Netherlands and Belgium all co-operate in defending a crucial sector of the Central Front. I am concerned at the steady erosion of Belgian capabilities as a result of the long decline in your country's defence spending. If Europe is to have a common security identity, then it is right that all of its members should make a serious effort, commensurate with their economic resources. I hope that in due course Spain too will feel able to look at ways in which its forces can contribute to a European, and not just a Spanish, defence posture.

We also need, all of us, better value for money. Much has been done by way of transferring tail to tooth. In Britain we have far leaner-looking armed forces today than a few years ago. By transferring 4,000 army personnel from support to combat roles we have been able to form a twelfth armoured regiment in BAOR and re-equip three additional regiments with *Challenger* tanks. In the Royal Air Force we have been able to increase our front line strength by 13% with no net increase in manpower.

But we must see what more can be done in the fields of economic management and defence procurement, and what more can be done to improve the management of collaborative projects. In particular, what lessons can be learned from past co-operation and the numbers of partners engaged in particular projects? How can we collaborate better in the field of military research? Should we not take a fresh look at role specialisation, at more joint training, and at the scope for establishing more common programmes on the NATO/AWACS model? How can we make the most of the developing role of the Independent European Programme Group? How else can we improve value for money?

European concern over its own defence identity is scarely a new idea. The main organisation left from those early years is the Western European Union [WEU]. It became moribund for some years. But it has recently been revived. It is instructive to look at the reasons. There has been a growing perception that a European forum *was* needed in which we Europeans could consult one another about our common fundamental security needs. At the same time the European Community's political co-operation framework has not developed a full security dimension.

Yet recent developments, from the Reykjavik Summit to the prospect of an INF agreement, underline the need for the European countries to consult more closely among themselves about their defence interests as well as with the US.

A Europe which gets its own ideas straight is a far more rewarding partner for the United States; and far more likely to have its views taken seriously than a Europe which speaks with a multitude of voices. If we want our particular European concerns to be clearly perceived and taken into account in negotiations between the United States and the Soviet Union, then we must argue them out clearly among ourselves and come wherever possible to a common view.

That is what we have been attempting, increasingly, to do within the WEU. We have tried to make it into a forum where Foreign and Defence Ministers can talk through the problems of European defence, as a prelude to bringing a clearer and distinctive European contribution into the deliberations of the Alliance as a whole.

There are traps which we must avoid. We don't want the WEU unnecessarily to duplicate work which is done just as well or better elsewhere. We don't want it to upset our other partners in the Alliance. We don't want it to become obsessed with institutional problems at the expense of the practical contribution, which is the vital one. The NATO Alliance itself must remain *the* decision-making forum for defence matters. A more effective WEU must bring more, not less, strength to the Alliance.

I am confident that these pitfalls can be avoided and that the WEU can become a more relevant, and above all a more useful body. We have already done away with the anachronistic protocols and arms-control mechanisms. If the organisation can be made more effective by concentrating its activities in one place—close to NATO here in Brussels—then we should consider that, provided that this is not an excuse for creating a new bureaucracy.

The advantages of broadening the WEU's membership need to be balanced against the importance of maintaining its sense of cohesion and purpose. The present membership is homogenous. We, you, your Benelux neighbours, France, Germany and Italy all see broadly eye-to-eye on both the nuclear and the conventional aspects of defence. The test of membership must be readiness to accept and implement in full the commitments of the Treaty. And to make—all of us—the practical military dispositions to give effect to this.

But what our people want above all is to get the best possible value out of the European contribution to the Alliance. That means that the accent of the WEU's activities must be on the practical, not the theoretical. What matters is not the number of meetings that we hold or the quantity of paper we produce, but the quality of the distinctive contribution we can make to the thinking and effectiveness of our Alliance. Alongside a clearer view of the needs of European security, we must make real headway in such things as the standardisation of equipment, the elimination of waste, and the rationalisation of infrastructure and support services. The effort of breathing new life into the WEU must be justified by the extent to which it succeeds in all these ways in strengthening the European contribution to the NATO Alliance as a whole.

In conclusion, let me stress that the Atlantic Alliance is and must remain the ultimate guarantor of European security. But if it is to remain in good shape it must adapt to changing circumstances. The management of differences of perception between the US and Europe will be crucial to this process. A better European defence effort galvanised perhaps through the WEU, can lead to a more substantial European pillar of the Alliance. It can help the United States to justify to sceptics the maintenance of its own contribution to European defence.

But to achieve this we Europeans must do more than just talk. If we want to be treated as equals within NATO, and as serious interlocutors by the Soviet Union, we must not only have our own ideas. We must show that we are willing to take the hard decisions and fulfil the commitments necessary to take proper responsibility for our defence. That is in the interest of Europe. It is in the interest of the United States. And it is in the best interest of the Alliance which guarantees peace with freedom for the great democracies on both sides of the Atlantic.

Arms Control: The Intermediate-Range Nuclear Forces Treaty

TREATY BETWEEN THE USA AND USSR ON ELIMINATION OF THEIR INTERMEDIATE-RANGE AND SHORTER-RANGE MISSILES, DECEMBER 8, 1987 (EXCERPTS)[1]

The United States of America and the Union of Soviet Socialist Republics, hereinafter referred to as the Parties,

Conscious that nuclear war would have devastating consequences for all mankind,

Guided by the objective of strengthening strategic stability,

Convinced that the measures set forth in this Treaty will help to reduce the risk of outbreak of war and strengthen international peace and security, and

Mindful of their obligations under Article VI of the Treaty on the Non-Proliferation of Nuclear Weapons,

Have agreed as follows:

Article I

In accordance with the provisions of this Treaty which includes the Memorandum of Understanding and Protocols which form an integral part thereof, each Party shall eliminate its intermediate-range and shorter-range missiles, not have such systems thereafter, and carry out the other obligations set forth in this Treaty.

Article II

For the purposes of this Treaty:

. . .(5) The term "intermediate-range missile" means a GLBM [ground-launched ballistic

---

[1]*Survival* (March–April 1988), pp. 63–80.

missile] or a GLCM [ground-launched cruise missile], having a range capability in excess of 1000 kilometers but not in excess of 5500 kilometers.
(6) The term "shorter-range missile" means a GLBM or a GLCM having a range capability equal to or in excess of 500 kilometers but not in excess of 1000 kilometers. . . .

Article III

(1) For the purposes of this Treaty, existing types of intermediate-range missiles are:

(*a*) for the USA, missiles of the types designated by the USA as the *Pershing* II and the BGM-109G, which are known to the USSR by the same designations; and

(*b*) for the USSR, missiles of the types designated by the USSR as the RSD-10, the R-12 and the R-14, which are known to the USA as the SS-20, the SS-4 and the SS-5, respectively.

(2) For the purposes of this Treaty, existing types of shorter-range missiles are:

(*a*) for the USA, missiles of the type designated by the USA as the *Pershing* IA, which is known to the USSR by the same designation; and

(*b*) for the USSR, missiles of the types designated by the USSR as the OTR-22 and the OTR-23, which are known to the USA as the SS-12 and the SS-23, respectively.

Article IV

(1) Each Party shall eliminate all its intermediate-range missiles and launchers of such missiles, and all support structures and support equipment of the categories listed in the Memorandum of Understanding associated with such missiles and launchers, so that no later than three years after entry into force of this Treaty and thereafter no such missiles, launchers, support structures or support equipment shall be possessed by either Party.
(2) To implement paragraph I of this Article, upon entry into force of this Treaty, both Parties shall begin and continue throughout the duration of each phase, the reduction of all types of their deployed and non-deployed intermediate-range missiles and deployed and non-deployed launchers of such missiles and support structures and support equipment associated with such missiles and launchers in accordance with the provisions of this Treaty. These reductions shall be implemented in two phases so that:

(*a*) by the end of the first phase, that is, no later than 29 months after entry into force of this Treaty:

(*i*) the number of deployed launchers of intermediate-range missiles for each Party shall not exceed the number of launchers that are capable of carrying or containing at one time missiles considered by the Parties to carry 171 warheads;

(*ii*) the number of deployed intermediate-range missiles for each Party shall not exceed the number of such missiles considered by the Parties to carry 180 warheads;

(*iii*) the aggregate number of deployed and non-deployed launchers of intermediate-range missiles for each Party shall not exceed the number of launchers that are capable of carrying or containing at one time missiles considered by the Parties to carry 200 warheads;

(*iv*) the aggregate number of deployed and non-deployed intermediate-range missiles for each Party shall not exceed the number of such missiles considered by the Parties to carry 200 warheads; and

(*v*) the ratio of the aggregate number of deployed and non-deployed intermediate-range GLBM of existing types for each Party to the aggregate number of deployed and non-deployed intermediate-range missiles of existing types possessed by that Party shall not exceed the ratio of such intermediate-range GLBM to such intermediate-range missiles for the Party as of November 1, 1987, as set forth in the Memorandum of Understanding; and

(*b*) by the end of the second phase, that is, no later than three years after entry into force of this Treaty, all intermediate-range missiles of each Party, launchers of such missiles and all support structures and support equipment of the categories listed in the Memorandum of Understanding associated with such missiles and launchers, shall be eliminated.

Article V

(1) Each Party shall eliminate all its shorter-range missiles and launchers of such missiles, and all support equipment of the categories listed in the Memorandum of Understanding associated with such missiles and launchers, so that no later than 18 months after entry into force of this Treaty and thereafter no such missiles, launchers or support equipment shall be possessed by either Party.

(2) No later than 90 days after entry into force of this Treaty, each Party shall complete the removal of all its deployed shorter-range missiles and deployed and non-deployed launchers of such missiles to elimination facilities and shall retain them at those locations until they are eliminated in accordance with the procedures set forth in the Protocol on Elimination. No later than 12 months after entry into force of this Treaty, each Party shall complete the removal of all its non-deployed shorter-range missiles to elimination facilities and shall retain them at those locations until they are eliminated in accordance with the procedures set forth in the Protocol on Elimination.

(3) Shorter-range missiles and launchers of such missiles shall not be located at the same elimination facility. Such facilities shall be separated by no less than 1000 kilometers.

Article VI

(1) Upon entry into force of this Treaty and thereafter, neither Party shall:

(*a*) produce or flight-test any intermediate-range missiles or produce any stages of such missiles or any launchers of such missiles; or

(*b*) produce, flight-test or launch any shorter-range missiles or produce any stages of such missiles or any launchers of such missiles.

(2) Notwithstanding paragraph 1 of this Article, each Party shall have the right to produce a type of GLBM not limited by this Treaty which uses a stage which is outwardly similar to, but not interchangeable with, a stage of an existing type of intermediate-range GLBM having more than one stage, providing that that Party does not produce any other stage which is outwardly similar to, but not interchangeable with, any other stage of an existing type of intermediate-range GLBM.

Article VII

For the purposes of this Treaty:

(1) If a ballistic missile or a cruise missile has been flight-tested or deployed for weapon delivery, all missiles of that type shall be considered to be weapon-delivery vehicles. . . .

Article VIII

(1) All intermediate-range missiles and launchers of such missiles shall be located in deployment areas, at missile support facilities or shall be in transit. Intermediate-range missiles or launchers of such missiles shall not be located elsewhere. . . .

Article IX

(1) The Memorandum of Understanding contains categories of data relevant to obligations undertaken with regard to this Treaty and lists all intermediate-range and shorter-range missiles, launchers of such missiles, and support structures and support equipment associated with such missiles and launchers, possessed by the Parties as of November 1, 1987. Updates of that data and notifications required by this Article shall be provided according to the categories of data contained in the Memorandum of Understanding.
(2) The Parties shall update that data and provide the notifications required by this Treaty through the Nuclear Risk Reduction Centers, established pursuant to the Agreement Between the United States of America and the Union of Soviet Socialist Republics on the Establishment of Nuclear Risk Reduction Centers of September 15, 1987. . . .
(5) Upon entry into force of this Treaty and thereafter, each Party shall provide the following notifications to the other Party:

(*a*) notification, no less than 30 days in advance, of the scheduled date of the elimination of a specific deployment area, missile operating base or missile support facility;

(*b*) notification, no less than 30 days in advance, of changes in the number or location of elimination facilities, including the location and scheduled date of each change;

(*c*) notification, except with respect to launches of intermediate-range missiles for the purpose of their elimination, no less than 30 days in advance, of the scheduled date of the initiation of the elimination of intermediate-range and shorter-range missiles, and stages of such missiles, and launchers of such missiles and support structures and support equipment associated with such missiles and launchers, including:

(*i*) the number and type of items of missile systems to be eliminated;

(*ii*) the elimination site;

(*iii*) for intermediate-range missiles, the location from which such missiles, launchers of such missiles and support equipment associated with such missiles and launchers are moved to the elimination facility; and

(*iv*) except in the case of support structures, the point of entry to be used by an inspection team conducting an inspection pursuant to paragraph 7 of Article XI of this Treaty and the estimated time of departure of an inspection team from the point of entry to the elimination facility;

(*d*) notification, no less than ten days in advance, of the scheduled date of the launch, or the scheduled date of the initiation of a series of launches, of intermediate-range missiles for the purpose of their elimination, including:

(*i*) the type of missiles to be eliminated;

(*ii*) the location of the launch, or, if elimination is by a series of launches, the location of such launches and the number of launches in the series;

(*iii*) the point of entry to be used by an inspection team conducting an inspection pursuant to paragraph 7 of Article XI of this Treaty; and

(*iv*) the estimated time of departure of an inspection team from the point of entry to the elimination facility;

(*e*) notification, no later than 48 hours after they occur, of changes in the number of intermediate-range and shorter-range missiles, launchers of such missiles and support structures and support equipment associated with such missiles and launchers resulting from elimination as described in the Protocol on Elimination, including:

(*i*) the number and type of items of a missile system which were eliminated; and

(*ii*) the date and location of such elimination; and

(*f*) notification of transit of intermediate-range or shorter-range missiles or launchers of such missiles, or the movement of training missiles or training launchers for such intermediate-range and shorter-range missiles, no later than 48 hours after it has been completed, including:

(*i*) the number of missiles or launchers;

(*ii*) the points, dates and times of departure and arrival;

(*iii*) the mode of transport; and

(*iv*) the location and time at that location at least once every four days during the period of transit.

(6) Upon entry into force of this Treaty and thereafter, each Party shall notify the other Party, no less than ten days in advance, of the scheduled date and location of the launch of a research and development booster system as described in paragraph 12 of Article VII of this Treaty.

Article X

. . .(5) Each Party shall have the right, during the first six months after entry into force of this Treaty, to eliminate by means of launching no more than 100 of its intermediate-range missiles.

(6) Intermediate-range and shorter-range missiles which have been tested prior to entry into force of this Treaty, but never deployed, and which are not existing types of intermediate-range or shorter-range missiles listed in Article III of this Treaty, and launchers of such missiles, shall be eliminated within six months after entry into force of this Treaty in accordance with the procedures set forth in the Protocol on Elimination. Such missiles are:

(*a*) for the USA, missiles of the type designated by the USA as the *Pershing* IB, which is known to the USSR by the same designation; and

(*b*) for the USSR, missiles of the type designated by the USSR as the RK-55, which is known to the USA as the SSC-X-4. . . .

Article XI

(1) For the purpose of ensuring verification of compliance with the provisions of this Treaty, each Party shall have the right to conduct on-site inspections. The Parties shall implement on-site inspections in accordance with this Article, the Protocol on inspection and the Protocol on Elimination. . . .

(3) Beginning 30 days after entry into force of this Treaty, each Party shall have the right to conduct inspections at all missile operating bases and missile support facilities specified in the Memorandum of Understanding other than missile production facilities, and at all elimination facilities included in the initial data update required by paragraph 3 of Article IX of this Treaty. These inspections shall be completed no later than 90 days after entry into force of this Treaty. The purpose of these inspections shall be to verify the number of missiles, launchers, support structures and support equipment and other data, as of the date of entry into force of this Treaty, provided pursuant to paragraph 3 of Article IX of this Treaty.

(4) Each Party shall have the right to conduct inspections to verify the elimination, notified pursuant to paragraph 5(*a*) of Article IX of this Treaty, of missile operating bases and missile support facilities other than missile production facilities, which are thus no longer subject to inspections pursuant to paragraph 5(*a*) of this Article. Such an inspection shall be carried out within 60 days after the scheduled date of the elimination of that facility. If a Party conducts an inspection at a particular facility pursuant to paragraph 3 of this Article after the scheduled

date of the elimination of that facility, then no additional inspection of that facility pursuant to this paragraph shall be permitted.

(5) Each Party shall have the right to conduct inspections pursuant to this paragraph for 13 years after entry into force of this Treaty. Each Party shall have the right to conduct 20 such inspections per calendar year during the first three years after entry into force of this Treaty, 15 such inspections per calendar year during the subsequent five years, and ten such inspections per calendar year during the last five years. Neither Party shall use more than half of its total number of these inspections per calendar year within the territory of any one basing country. Each Party shall have the right to conduct:

(*a*) inspections, beginning 90 days after entry into force of this Treaty, of missile operating bases and missile support facilities other than elimination facilities and missile production facilities, to ascertain, according to the categories of data specified in the Memorandum of Understanding, the numbers of missiles, launchers, support structures and support equipment located at each missile operating base or missile support facility at the time of the inspection; and

(*b*) inspections of former missile operating bases and former missile support facilities eliminated pursuant to paragraph 8 of Article X of this Treaty other than former missile production facilities.

(6) Beginning 30 days after entry into force of this Treaty, each Party shall have the right, for 13 years after entry into force of this Treaty, to inspect by means of continuous monitoring:

(*a*) the portals of any facility of the other Party at which the final assembly of a GLBM using stages, any of which is outwardly similar to a stage of a solid-propellant GLBM listed in Article III of this Treaty, is accomplished; or

(*b*) if a Party has no such facility, the portals of an agreed former missile production facility at which existing types of intermediate-range or shorter-range GLBM were produced.

The Party whose facility is to be inspected pursuant to this paragraph shall ensure that the other Party is able to establish a permanent continuous monitoring system at that facility within six months after entry into force of this Treaty or within six months of initiation of the process of final assembly described in subparagraph (*a*). If, after the end of the second year after entry into force of this Treaty, neither Party conducts the process of final assembly described in subparagraph (*a*) for a period of 12 consecutive months, then neither Party shall have the right to inspect by means of continuous monitoring any missile production facility of the other Party unless the process of final assembly as described in subparagraph (*a*) is initiated again. Upon entry into force of this Treaty, the facilities to be inspected by continuous monitoring shall be: in accordance with subparagraph (*b*), for the USA, Hercules Plant Number 1, at Magna, Utah; in accordance with subparagraph (*a*), for the USSR, the Votkinsk Machine Building Plant, Udmurt Autonomous Soviet Socialist Republic, Russian Soviet Federative Socialist Republic.

(7) Each Party shall conduct inspections of the process of elimination, including elimination of intermediate-range missiles by means of launching, of intermediate-range and shorter-range missiles and launchers of such missiles and support equipment associated with such missiles and launchers carried out at elimination facilities in accordance with Article X of this Treaty and the Protocol on Elimination. Inspectors conducting inspections provided for in this paragraph shall determine that the processes specified for the elimination of the missiles, launchers and support equipment have been completed.

(8) Each Party shall have the right to conduct inspections to confirm the completion of the process of elimination of intermediate-range and shorter-range missiles and launchers of such missiles and support equipment associated with such missiles and launchers eliminated pursuant to Section V of the Protocol on Elimination, and of training missiles, training

missile stages, training launch canisters and training launchers eliminated pursuant to Sections II, IV and V of the Protocol on Elimination.

Article XII

(1) For the purpose of ensuring verification of compliance with the provisions of this Treaty, each Party shall use national technical means of verification at its disposal in a manner consistent with generally recognized principles of international law.
(2) Neither Party shall:

(*a*) interfere with national technical means of verification of the other Party operating in accordance with paragraph 1 of this Article; or

(*b*) use concealment measures which impede verification of compliance with the provisions of this Treaty by national technical means of verification carried out in accordance with paragraph 1 of this Article. This obligation does not apply to cover or concealment practices, within a deployment area, associated with normal training, maintenance and operations, including the use of environmental shelters to protect missiles and launchers.
(3) To enhance observation by national technical means of verification, each Party shall have the right until a treaty between the Parties reducing and limiting strategic offensive arms enters into force, but in any event for no more than three years after entry into force of this Treaty, to request the implementation of cooperative measures at deployment bases for road-mobile GLBM with a range capability in excess of 5500 kilometers, which are not former missile operating bases eliminated pursuant to paragraph 8 of Article X of this Treaty. The Party making such a request shall inform the other Party of the deployment base at which cooperative measures shall be implemented. The Party whose base is to be observed shall carry out the following cooperative measures:

(*a*) no later than six hours after such a request, the Party shall have opened the roofs of all fixed structures for launchers located at the bases, removed completely all missiles on launchers from such fixed structures for launchers and displayed such missiles on launchers in the open without using concealment measures; and

(*b*) the Party shall leave the roofs open and the missiles on launchers in place until twelve hours have elapsed from the time of the receipt of a request for such an observation.

Each Party shall have the right to make six such requests per calendar year. Only one deployment base shall be subject to these cooperative measures at any one time.

Article XIII

(1) To promote the objectives and implementation of the provisions of this Treaty, the Parties hereby establish the Special Verification Commission. The Parties agree that, if either Party so request, they shall meet within the framework of the Special Verification Commission to:

(*a*) resolve questions relating to compliance with the obligations assumed; and

(*b*) agree upon such measures as may be necessary to improve the viability and effectiveness of this Treaty.
(2) The Parties shall use the Nuclear Risk Reduction Centers, which provide for continuous communication between the Parties, to:

(*a*) exchange data and provide notifications as required by paragraphs 3, 4, 5 and 6 of Article IX of this Treaty and the Protocol on Elimination;

(*b*) provide and receive the information required by paragraph 9 of Article X of this Treaty;

(*c*) provide and receive notifications of inspections as required by Article XI of this Treaty and the Protocol on Inspection; and

(*d*) provide and receive requests for cooperative measures as provided for in paragraph 3 of Article XII of this Treaty.

Article XIV

The Parties shall comply with this Treaty and shall not assume any international obligations or undertakings which would conflict with its provisions.

Article XV

(1) This Treaty shall be of unlimited duration.

(2) Each Party shall, in exercising its national sovereignty, have the right to withdraw from this Treaty if it decides that extraordinary events related to the subject matter of this Treaty have jeopardized its supreme interests. It shall give notice of its decision to withdraw to the other Party six months prior to withdrawal from this Treaty. Such notice shall include a

Summary of the Data Contained in the INF Memorandum of Understanding

	Launchers			Missiles		
	Deployed	Non-deployed	Total	Deployed	Non-deployed	Total
USA						
LRINF						
Pershing II	115	51	166	120	127	247
GLCM	99	17	116	309	133	442
Subtotal	214	68	282	429	260	689
SRINF						
Pershing IA	—	1	1	—	178	178
Total INF	214	69	283	429	438	867
USSR						
LRINF						
SS-20	405	118	523	405	245	650
SS-4	79	6	85	65	105	170
SS-5	—	—	—	—	6	6
SSC-X-4	—	6	6	—	84	84
Subtotal	484	130	614	470	440	910
SRINF						
SS-12	115	20	135	220	506	726
SS-23	82	20	102	167	33	200
Subtotal	197	40	237	387	539	926
Total INF	681	170	851	857	979	1,836

Source: Compiled by IISS from INF MOU.

statement of the extraordinary events the notifying Party regards as having jeopardized its supreme interests.

Article XVI

Each Party may propose amendments to this Treaty. Agreed amendments shall enter into force in accordance with the procedures set forth in Article XVII governing the entry into force of this Treaty.

Article XVII

(1) This Treaty, including the Memorandum of Understanding and Protocols, which form an integral part thereof, shall be subject to ratification in accordance with the constitutional procedures of each Party. This Treaty shall enter into force on the date of the exchange of instruments of ratification.
(2) This Treaty shall be registered pursuant to Article 102 of the Charter of the United Nations.

Done at Washington on December 8, 1987, in two copies, each in the English and Russian languages, both texts being equally authentic.
For the United States of America:
President of the United States of America
For the Union of Soviet Socialist Republics:
General Secretary of the Central Committee of the CPSU

Protocol on Procedures Governing the Elimination of the Missile Systems Subject to the Treaty

. . .II. Procedures for Elimination at Elimination Facilities

(1) In order to ensure the reliable determination of the type and number of missiles, missile stages, front sections, launch canisters, launchers, missile transporter vehicles, missile erectors and launch stands, as well as training missiles, training missile stages, training launch canisters and training launchers, indicated in Section I of this Protocol, being eliminated at elimination facilities, and to preclude the possibility of restoration of such items for purposes inconsistent with the provisions of the Treaty, the Parties shall fulfill the requirements below.
(2) The conduct of the elimination procedures for the items of missile systems listed in paragraph 1 of this Section, except for training missiles, training missile stages, training launch canisters and training launchers, shall be subject to on-site inspection in accordance with Article XI of the Treaty and the Protocol on Inspection. The Parties shall have the right to conduct on-site inspections to confirm the completion of the elimination procedures set forth in paragraph 11 of this Section for training missiles, training missile stages, training launch canisters and training launchers. The Party possessing such a training missile, training missile stage, training launch canister or training launcher shall inform the other Party of the name and coordinates of the elimination facility at which the on-site inspection may be conducted as well as the date on which it may be conducted. Such information shall be provided no less than 30 days in advance of that date.
(3) Prior to a missile's arrival at the elimination facility, its nuclear warhead device and guidance elements may be removed. . . .

(10) The specific procedures for the elimination of the items of missile systems listed in paragraph 1 of this Section shall be as follows, unless the Parties agree upon different procedures to achieve the same result as the procedures identified in this paragraph:

For the Pershing *II:*

MISSILE:

(*a*) missile stages shall be eliminated by explosive demolition or burning;

(*b*) solid fuel, rocket nozzles and motor cases not destroyed in this process shall be burned, crushed, flattened or destroyed by explosion; and

(*c*) front section, minus nuclear warhead device and guidance elements, shall be crushed or flattened.

LAUNCHER:

(*a*) erector-launcher mechanism shall be removed from launcher chassis;

(*b*) all components of erector-launcher mechanism shall be cut at locations that are not assembly joints into two pieces of approximately equal size;

(*c*) missile launch support equipment, including external instrumentation compartments, shall be removed from launcher chassis; and

(*d*) launcher chassis shall be cut at a location that is not an assembly joint into two pieces of approximately equal size.

For the BGM-109G:

MISSILE:

(*a*) missile airframe shall be cut longitudinally into two pieces;

(*b*) wings and tail section shall be severed from missile airframe at locations that are not assembly joints; and

(*c*) front section, minus nuclear warhead device and guidance elements, shall be crushed or flattened.

LAUNCH CANISTER:

Launch canister shall be crushed, flattened, cut into two pieces of approximately equal size or destroyed by explosion.

LAUNCHER:

(*a*) erector-launcher mechanism shall be removed from launcher chassis;

(*b*) all components of erector-launcher mechanism shall be cut at locations that are not assembly joints into two pieces of approximately equal size;

(*c*) missile launch support equipment, including external instrumentation compartments, shall be removed from launcher chassis; and

(*d*) launcher chassis shall be cut at a location that is not an assembly joint into two pieces of approximately equal size.

For the Pershing *IA:*

MISSILE:

(*a*) missile stages shall be eliminated by explosive demolition or burning;

(*b*) solid fuel, rocket nozzles and motor cases not destroyed in this process shall be burned, crushed, flattened or destroyed by explosion; and

(*c*) front section, minus nuclear warhead device and guidance elements, shall be crushed or flattened.

LAUNCHER:

(*a*) erector-launcher mechanism shall be removed from launcher chassis;

(*b*) all components of erector-launcher mechanism shall be cut at locations that are not assembly joints into two pieces of approximately equal size;
(*c*) missile launch support equipment, including external instrumentation compartments, shall be removed from launcher chassis; and
(*d*) launcher chassis shall be cut at a location that is not an assembly joint into two pieces of approximately equal size.

For the Pershing *IB:*

MISSILE:
(*a*) missile stage shall be eliminated by explosive demolition or burning;
(*b*) solid fuel, rocket nozzle and motor case not destroyed in this process shall be burned, crushed, flattened or destroyed by explosion; and
(*c*) front section, minus nuclear warhead device and guidance elements, shall be crushed or flattened.

For the SS-20

MISSILE:
(*a*) missile shall be eliminated by explosive demolition of the missile in its launch canister or by burning missile stages;
(*b*) solid fuel, rocket nozzles and motor cases not destroyed in this process shall be burned, crushed, flattened or destroyed by explosion; and
(*c*) front section, including reentry vehicles, minus nuclear warhead devices, and instrumentation compartment, minus guidance elements, shall be crushed or flattened.
LAUNCH CANISTER:

Launch canister shall be destroyed by explosive demolition together with a missile, or shall be destroyed separately by explosion, cut into two pieces of approximately equal size, crushed or flattened.

LAUNCHER:
(*a*) erector-launcher mechanism shall be removed from launcher chassis;
(*b*) all components of erector-launcher mechanism shall be cut at locations that are not assembly joints into two pieces of approximately equal size;
(*c*) missile launch support equipment, including external instrumentation compartments, shall be removed from launcher chassis;
(*d*) mountings of erector-launcher mechanism and launcher leveling supports shall be cut off launcher chassis;
(*e*) launcher leveling supports shall be cut at locations that are not assembly joints into two pieces of approximately equal size; and
(*f*) a portion of the launcher chassis, at least 0.78 meters in length, shall be cut off aft of the rear axle.
MISSILE TRANSPORTER VEHICLE:
(*a*) all mechanisms associated with missile loading and mounting shall be removed from transporter vehicle chassis;
(*b*) all mountings of such mechanisms shall be cut off transporter vehicle chassis;
(*c*) all components of the mechanisms associated with missile loading and mounting shall be cut at locations that are not assembly joints into two pieces of approximately equal size;
(*d*) external instrumentation compartments shall be removed from transporter vehicle chassis;
(*e*) transporter vehicle leveling supports shall be cut off transporter vehicle chassis and cut at

locations that are not assembly joints into two pieces of approximately equal size; and (*f*) a portion of the transporter vehicle chassis, at least 0.78 meters in length, shall be cut off aft of the rear axle. . . .

Protocol Regarding Inspections Relating to the Treaty

. . .III. Pre-Inspection Requirements

(1) Inspections to ensure verification of compliance by the Parties with the obligations assumed under the Treaty shall be carried out by inspectors designated in accordance with paragraphs 3 and 4 of this Section.
(2) No later than one day after entry into force of the Treaty, each Party shall provide to the other Party: a list of its proposed aircrew members; a list of its proposed inspectors who will carry out inspections pursuant to paragraphs 3, 4, 5, 7 and 8 of Article XI of the Treaty; and a list of its proposed inspectors who will carry out inspection activities pursuant to paragraph 6 of Article XI of the Treaty. None of these lists shall contain at any time more than 200 individuals.
(3) Each Party shall review the lists of inspectors and aircrew members proposed by the other Party. With respect to an individual included on the list of proposed inspectors who will carry out inspection activities pursuant to paragraph 6 of Article XI of the Treaty, if such an individual is unacceptable to the Party reviewing the list, that Party shall, within 20 days, so inform the Party providing the list, and the individual shall be deemed not accepted and shall be deleted from the list. With respect to an individual on the list of proposed aircrew members or the list of proposed inspectors who will carry out inspections pursuant to paragraphs 3, 4, 5, 7 and 8 of Article XI of the Treaty, each Party, within 20 days after the receipt of such lists, shall inform the other Party of its agreement to the designation of each inspector and aircrew member proposed. Inspectors shall be citizens of the inspecting Party.
(4) Each Party shall have the right to amend its lists of inspectors and aircrew members. New inspectors and aircrew members shall be designated in the same manner as set forth in paragraph 3 of this Section with respect to the initial lists.
(5) Within 30 days of receipt of the initial lists of inspectors and aircrew members, or of subsequent changes thereto, the Party receiving such information shall provide, or shall ensure the provision of, such visas and other documents to each individual to whom it has agreed as may be required to ensure that each inspector or aircrew member may enter and remain in the territory of the Party or basing country in which an inspection site is located throughout the in-country period for the purpose of carrying out inspection activities in accordance with the provisions of this Protocol. Such visas and documents shall be valid for a period of at least 24 months.
(6) To exercise their functions effectively, inspectors and aircrew members shall be accorded, throughout the in-country period, privileges and immunities in the country of the inspection site as set forth in the Annex to this Protocol.
(7) Without prejudice to their privileges and immunities, inspectors and aircrew members shall be obliged to respect the laws and regulations of the State on whose territory an inspection is carried out and shall be obliged not to interfere in the internal affairs of that State. In the event the inspected Party determines that an inspector or aircrew member of the other Party has violated the conditions governing inspection activities set forth in this Protocol, or has ever committed a criminal offense on the territory of the inspected Party or a basing country, or has ever been sentenced for committing a criminal offense or expelled by the inspected Party or a basing country, the inspected Party making such a determination shall

so notify the inspecting Party, which shall immediately strike the individual from the lists of inspectors or the list of aircrew members. If, at that time, the individual is on the territory of the inspected Party or a basing country, the inspecting Party shall immediately remove that individual from the country.
(8) Within 30 days after entry into force of the Treaty, each Party shall inform the other Party of the standing diplomatic clearance number for airplanes of the Party transporting inspectors and equipment necessary for inspection into and out of the territory of the Party or basing country in which an inspection site is located. Aircraft routings to and from the designated point of entry shall be along established international airways that are agreed upon by the Parties as the basis for such diplomatic clearance.

IV. Notifications

(1) Notification of an intention to conduct an inspection shall be made through the Nuclear Risk Reduction Centers. The receipt of this notification shall be acknowledged through the Nuclear Risk Reduction Centers by the inspected Party within one hour of its receipt.

(*a*) For inspections conducted pursuant to paragraphs 3, 4 or 5 of Article XI of the Treaty, such notifications shall be made no less than 16 hours in advance of the estimated time of arrival of the inspection team at the point of entry and shall include:

(*i*) the point of entry;

(*ii*) the date and estimated time of arrival at the point of entry;

(*iii*) the date and time when the specification of the inspection site will be provided; and

(*iv*) the names of inspectors and aircrew members.

(*b*) For inspections conducted pursuant to paragraphs 7 or 8 of Article XI of the Treaty, such notifications shall be made no less than 72 hours in advance of the estimated time of arrival of the inspection team at the point of entry and shall include:

(*i*) the point of entry;

(*ii*) the date and estimated time of arrival at the point of entry;

(*iii*) the site to be inspected and the type of inspection; and

(*iv*) the names of inspectors and aircrew members.

(2) The date and time of the specification of the inspection site as notified pursuant to paragraph 1(*a*) of this Section shall fall within the following time intervals:

(*a*) for inspections conducted pursuant to paragraphs 4 or 5 of Article XI of the Treaty, neither less than four hours nor more than 24 hours after the estimated date and time of arrival at the point of entry; and

(*b*) for inspections conducted pursuant to paragraph 3 of Article XI of the Treaty, neither less than four hours nor more than 48 hours after the estimated date and time of arrival at the point of entry.

(3) The inspecting Party shall provide the inspected Party with a flight plan, through the Nuclear Risk Reduction Centers, for its flight from the last airfield prior to entering the airspace of the country in which the inspection site is located to the point of entry, no less than six hours before the scheduled departure time from that airfield. Such a plan shall be filed in accordance with the procedures of the International Civil Aviation Organization applicable to civil aircraft. The inspecting Party shall include in the remarks section of each flight plan the standing diplomatic clearance number and the notation: "Inspection aircraft. Priority clearance processing required".

(4) No less than three hours prior to the scheduled departure of the inspection team from the last airfield prior to entering the airspace of the country in which the inspection is to take

place, the inspected Party shall ensure that the flight plan filed in accordance with paragraph 3 of this Section is approved so that the inspection team may arrive at the point of entry by the estimated arrival time.

(5) Either Party may change the point or points of entry to the territories of the countries within which its deployment areas, missile operating bases or missile support facilities are located, by giving notice of such change to the other Party. A change in a point of entry shall become effective five months after receipt of such notification by the other Party.

V. Activities Beginning Upon Arrival at the Point of Entry

(1) The in-country escort and a diplomatic aircrew escort accredited to the Government of either the inspected Party or the basing country in which the inspection site is located shall meet the inspection team and aircrew members at the point of entry as soon as the airplane of the inspecting Party lands. The number of aircrew members for each airplane shall not exceed ten. The in-country escort shall expedite the entry of the inspection team and aircrew, their baggage, and equipment and supplies necessary for inspection, into the country in which the inspection site is located. A diplomatic aircrew escort shall have the right to accompany and assist aircrew members throughout the in-country period. In the case of an inspection taking place on the territory of a basing country, the in-country escort may include representatives of that basing country. . . .

VII. Inspections Conducted Pursuant to Paragraphs 3, 4 or 5 of Article XI of the Treaty

. . .(7) Subject to the procedures of paragraphs 8 through 14 of this Section, inspectors shall have the right to inspect the entire inspection site, including the interior of structures, containers or vehicles, or including covered objects, whose dimensions are equal to or greater than the dimensions specified in Section VI of the Memorandum of Understanding for the missiles, stages of such missiles, launchers or support equipment of the inspected Party.

(8) A missile, a stage of such a missile or a launcher subject to the Treaty shall be subject to inspection only by external visual observation, including measuring, as necessary, the dimensions of such a missile, stage of such a missile or launcher. A container that the inspected Party declares to contain a missile or stage of a missile subject to the Treaty, and which is not sufficiently large to be capable of containing more than one missile or stage of such a missile of the inspected Party subject to the Treaty, shall be subject to inspection only by external visual observation, including measuring, as necessary, the dimensions of such a container to confirm that it cannot contain more than one missile or stage of such a missile of the inspected Party subject to the Treaty. Except as provided for in paragraph 14 of this Section, a container that is sufficiently large to contain a missile or stage of such a missile of the inspected Party subject to the Treaty that the inspected Party declares not to contain a missile or stage of such a missile subject to the Treaty shall be subject to inspection only by means of weighing or visual observation of the interior of the container, as necessary, to confirm that it does not, in fact, contain a missile or stage of such a missile of the inspected Party subject to the Treaty. If such a container is a launch canister associated with a type of missile not subject to the Treaty, and declared by the inspected Party to contain such a missile, it shall be subject to external inspection only, including use of radiation detection devices, visual observation and linear measurement, as necessary, of the dimensions of such a canister.

(9) A structure or container that is not sufficiently large to contain a missile, stage of such a missile or launcher of the inspected Party subject to the Treaty shall be subject to inspection

only by external visual observation including measuring, as necessary, the dimensions of such a structure or container to confirm that it is not sufficiently large to be capable of containing a missile, stage of such a missile or launcher of the inspected Party subject to the Treaty.

(10) Within a structure, a space which is sufficiently large to contain a missile, stage of such a missile or launcher of the inspected Party subject to the Treaty, but which is demonstrated to the satisfaction of the inspection team not to be accessible by the smallest missile, stage of a missile or launcher of the inspected Party subject to the Treaty shall not be subject to further inspection. If the inspected Party demonstrates to the satisfaction of the inspection team by means of a visual inspection of the interior of an enclosed space from its entrance that the enclosed space does not contain any missile, stage of such a missile or launcher of the inspected Party subject to the Treaty, such an enclosed space shall not be subject to further satisfaction of the inspection team by means of a visual inspection of the interior of an enclosed space from its entrance that the enclosed space does not contain any missile, stage of such a missile or launcher of the inspected Party subject to the Treaty, such an enclosed space shall not be subject to further inspection.

(11) The inspection team shall be permitted to patrol the perimeter of the inspection site and station inspectors at the exits of the site for the duration of the inspection.

(12) The inspection team shall be permitted to inspect any vehicle capable of carrying missiles, stages of such missiles, launchers or support equipment of the inspected Party subject to the Treaty at any time during the course of an inspection and no such vehicle shall leave the inspection site during the course of the inspection until inspected at site exits by the inspection team.

(13) Prior to inspection of a building within the inspection site, the inspection team may station subgroups at the exits of the building that are large enough to permit passage of any missile, stage of such a missile, launcher or support equipment of the inspected Party subject to the Treaty. During the time that the building is being inspected, no vehicle or object capable of containing any missile, stage of such a missile, launcher or support equipment of the inspected Party subject to the Treaty shall be permitted to leave the building until inspected.

(14) During an inspection conducted pursuant to paragraph 5(*b*) of Article XI of the Treaty, it shall be the responsibility of the inspected Party to demonstrate that a shrouded or environmentally protected object which is equal to or larger than the smallest missile, stage of a missile or launcher of the inspected Party subject to the Treaty is not, in fact, a missile, stage of such a missile or launcher of the inspected Party subject to the Treaty. This may be accomplished by partial removal of the shroud or environmental protection cover, measuring, or weighing the covered object or by other methods. If the inspected Party satisfies the inspection team by its demonstration that the object is not a missile, stage of such a missile or launcher of the inspected Party subject to the Treaty, then there shall be no further inspection of that object. If the container is a launch canister associated with a type of missile not subject to the Treaty, and declared by the inspected Party to contain such a missile, then it shall be subject to external inspection only, including use of radiation detection devices, visual observation and linear measurement, as necessary, of the dimensions of such a canister.

VIII. Inspections Conducted Pursuant to Paragraphs 7 or 8 of Article XI of the Treaty

(1) Inspections of the process of elimination of items of missile systems specified in the Protocol on Elimination carried out pursuant to paragraph 7 of Article XI of the Treaty shall

be conducted in accordance with the procedures set forth in this paragraph and the Protocol on Elimination.

(*a*) Upon arrival at the elimination facility, inspectors shall be provided with a schedule of elimination activities.

(*b*) Inspectors shall check the data which are specified in the notification provided by the inspected Party regarding the number and type of items of missile systems to be eliminated against the number and type of such items which are at the elimination facility prior to the initiation of the elimination procedures.

(*c*) Subject to paragraphs 3 and 11 of Section VI of this Protocol, inspectors shall observe the execution of the specific procedures for the elimination of the items of missile systems as provided for in the Protocol on Elimination. If any deviations from the agreed elimination procedures are found, the inspectors shall have the right to call the attention of the in-country escort to the need for strict compliance with the above-mentioned procedures. The completion of such procedures shall be confirmed in accordance with the procedures specified in the Protocol on Elimination.

(*d*) During the elimination of missiles by means of launching, the inspectors shall have the right to ascertain by visual observation that a missile prepared for launch is a missile of the type subject to elimination. The inspectors shall also be allowed to observe such a missile from a safe location specified by the inspected Party until the completion of its launch. During the inspection of a series of launches for the elimination of missiles by means of launching, the inspected Party shall determine the means of transport and route for the transportation of inspectors between inspection sites.

(2) Inspections of the elimination of items of missile systems specified in the Protocol on Elimination carried out pursuant to paragraph 8 of Article XI of the Treaty shall be conducted in accordance with the procedures set forth in Sections II, IV or V of the Protocol on Elimination or as otherwise agreed by the Parties.

IX. Inspection Activities Conducted Pursuant to Paragraph 6 of Article XI of the Treaty

(1) The inspected Party shall maintain an agreed perimeter around the periphery of the inspection site and shall designate a portal with not more than one rail line and one road which shall be within 50 meters of each other. All vehicles which can contain an intermediate-range GLBM or longest stage of such a GLBM of the inspected Party shall exit only through this portal.

(2) For the purposes of this Section, the provisions of paragraph 10 of Article VII of the Treaty shall be applied to intermediate-range GLBM of the inspected Party and the longest stage of such GLBM.

(3) There shall not be more than two other exits from the inspection site. Such exits shall be monitored by appropriate sensors. The perimeter of and exits from the inspection site may be monitored as provided for by paragraph 11 of Section VII of this Protocol.

(4) The inspecting Party shall have the right to establish continuous monitoring systems at the portal specified in paragraph 1 of this Section and appropriate sensors at the exits specified in paragraph 3 of this Section and carry out necessary engineering surveys, construction, repair and replacement of monitoring systems. . . .

X. Cancellation of Inspection

(1) An inspection shall be cancelled if, due to circumstances brought about by *force majeure*, it cannot be carried out. In the case of a delay that prevents an inspection team performing an inspection pursuant to paragraphs 3, 4 or 5 of Article XI of the Treaty, from arriving at the

inspection site during the time specified in paragraph 2 of Section VII of this Protocol, the inspecting Party may either cancel or carry out the inspection. If an inspection is cancelled due to circumstances brought about by *force majeure* or delay, then the number of inspections to which the inspecting Party is entitled shall not be reduced.

XI. Inspection Report

(1) For inspections conducted pursuant to paragraphs 3, 4, 5, 7 or 8 of Article XI of the Treaty, during post-inspection procedures, and no later than two hours after the inspection has been completed, the inspection team leader shall provide the in-country escort with a written inspection report in both the English and Russian languages. The report shall be factual. It shall include the type of inspection carried out, the inspection site, the number of missiles, stages of missiles, launchers and items of support equipment subject to the Treaty observed during the period of inspection and any measurements recorded pursuant to paragraph 10 of Section VI of this Protocol. Photographs taken during the inspection in accordance with agreed procedures, as well as the inspection site diagram provided for by paragraph 6 of Section VII of this Protocol, shall be attached to this report.
(2) For inspection activities conducted pursuant to paragraph 6 of Article XI of the Treaty, within 3 days after the end of each month, the inspection team leader shall provide the in-country escort with a written inspection report both in the English and Russian languages. The report shall be factual. It shall include the number of vehicles declared to contain a missile or stage of a missile as large or larger than and as heavy or heavier than an intermediate-range GLBM or longest stage of such a GLBM of the inspected Party that left the inspection site through the portal specified in paragraph 1 of Section IX of this Protocol during that month. The report shall also include any measurements of launch canisters or shipping containers contained in these vehicles recorded pursuant to paragraph 11 of Section VI of this Protocol. In the event the inspecting Party, under the provisions of paragraph 14(*c*) of Section IX of this Protocol, has viewed the interior of a launch canister or shipping container declared to contain a missile or stage of a missile as large or larger than and as heavy or heavier than an intermediate-range GLBM or longest stage of such a GLBM of the inspected Party, the report shall also include the measurements of the length and diameter of missile stages obtained during the inspection and recorded pursuant to paragraph 11 of Section VI of this Protocol. Photographs taken during the inspection in accordance with agreed procedures shall be attached to this report.
(3) The inspected Party shall have the right to include written comments in the report.
(4) The Parties shall, when possible, resolve ambiguities regarding factual information contained in the inspection report. Relevant clarifications shall be recorded in the report. The report shall be signed by the inspection team leader and by one of the members of the in-country escort. Each Party shall retain one copy of the report. . . .

ADDRESS BY FRG CHANCELLOR HELMUT SCHMIDT: STRATEGIC ARMS LIMITATION TALKS AND MUTUAL AND BALANCED FORCE REDUCTIONS TALKS, OCTOBER 28, 1977 (EXCERPT)[1]

Political and military balance is the prerequisite for our security, and I would warn against the illusion that there may be some ground or other for neglecting that balance. Indeed, it is not

[1]United States Arms Control and Disarmament Agency, *Documents on Disarmament, 1977* (Washington, D.C.: U.S. Government Printing Office, 1979), pp. 666–69.

only the prerequisite for our security but for fruitful progress with détente. In the first place we should recognize that—paradoxical as it may sound—a risky arms race and controlled arms limitations are closer together than they have ever been. It is only a narrow ridge which divides hope of peace and danger of war.

Secondly, changed strategic conditions have confronted us with new problems. SALT codifies the nuclear strategic balance between the United States and the Soviet Union. To put it differently: SALT neutralizes the strategic nuclear capabilities of the United States and the Soviet Union. In Europe this magnifies the significance of the disparities between East and West as regards nuclear tactical and conventional weapons.

Thirdly, we must press ahead with the Vienna negotiations on mutual balanced force reductions in order to make an important step in the direction of a better balance of power in Europe. No one can deny that the principle of parity is a sensible one. However, it must be the target of all arms limitation and control negotiations and must apply to all weapons. Neither side can accept to diminish its security unilaterally. We all have a vital interest in the negotiations between the two superpowers on the limitation and reduction of nuclear strategic weapons to continue and to lead to a reliable agreement. The nuclear powers have a special responsibility in this field.

On the other hand, especially we Europeans must be anxious to ensure that these negotiations do not neglect those factors which make up NATO's deterrence strategy. We are all faced with the dilemma of having to meet the moral and political necessity for arms limitation whilst, at the same time, maintaining a fully effective deterrent to war. We do not overlook the fact that both the United States and the Soviet Union must be equally anxious to remove the mutual strategic threat.

But strategic arms limitations confined to the United States and the Soviet Union would be bound to impair the security of the West European members of the alliance vis-à-vis Soviet military superiority in Europe if we do not succeed in removing the disparities in Europe parallel to the SALT negotiations. So long as this is not the case we must stick to a balance of all components of deterrence strategy. This means that the Alliance must be ready to make available enough of the right means to support the valid strategy and to prevent any developments that could remove the basis of our strategy, which is still the right one.

At the meeting of heads of state and government in 10 Downing Street last May, I said that the more we stabilize strategic nuclear parity, which my government has advocated from the outset, the more will it be necessary to achieve a conventional equilibrium as well. Today let me add, again in London: When SALT negotiations opened we Europeans did not have a clear enough view of the close connection between parity of strategic nuclear weapons on the one hand and tactical nuclear and conventional weapons on the other, or if we did, we did not articulate it clearly enough.

The aim now is to clearly see the connection between SALT and MBFR [Mutual and Balanced Force Reductions] and to draw the necessary practical conclusions. At that same top level meeting I said that there were in theory two possibilities of establishing a conventional balance with the Warsaw Pact states. One would be for the Western alliance to go in for a massive build-up of forces and weapons systems; the other for both NATO and the Warsaw Pact to reduce their force strength and achieve an overall balance at a lower level. I prefer the latter. The Vienna negotiations have not so far produced any results in terms of concrete wording. Since they began, the Warsaw Pact has increased the disparities as regards both conventional and tactical nuclear forces.

Up to now the Soviet Union has given no clear indication that Moscow is willing to accept the principle of parity for Europe and in this way make the principle of renunciation of force an element of the military balance as well. But until we see progress on MBFR, we shall have to abide by our principle of deterrence. And it is in this context and no other that all that has been said in the public discussion in all member states of the Western alliance on the subject of the neutron weapons are of value to the alliance as an additional element of the deterrence strategy, as a means of preventing war. But we should not limit ourselves to that examination. Beyond this we also should examine what relevance and weight this weapon has in regard to our efforts to achieve arms control.

This brings me back to the arms control negotiations. For the first time in history those negotiations are being conducted at a time when there exists a weapon capable of destroying all living things. Unlike former times, failure of such negotiations can no longer be compensated by banking on military victory. That is why it is of such crucial importance that all should realize the seriousness of the Vienna negotiations, and that is why results must be achieved there. In this respect I have seven "musts" or "must nots":

1. Both sides, all participants in the MBFR negotiations, must state their willingness to bring the negotiations to a positive conclusion and to be a party to reductions on a basis of equality.

2. Priority must be given to the aim—and it must be achieved without delay—of preventing any further increase in the military confrontation and thus dispelling apprehensions.

3. The threat of a surprise attack must be eliminated.

4. The confidence-building measures voluntarily agreed at the CSCE must be accepted with binding effect.

5. It must remain the principal objective of MBFR to achieve by means of reductions a balance of forces at a lower level.

6. Force reductions must be oriented to the principle of parity and must be verifiable. Parity and collectivity must be recognized as determining principles.

7. The capability of both alliance systems to organize their defence must not be impaired.

We should consider whether it is necessary to extend the confidence-building measures beyond the agreed scope. Because even if we ultimately achieve conventional parity within the MBFR reduction area we will still be considerably short of parity of conventional forces in Europe as a whole. And this all the more so as the Soviet Union has substantially increased the strategic transport capacities and thus could concentrate its forces outside the reduction area, whereas American forces would be cut off from Europe by the Atlantic.

COMMUNIQUÉ OF A SPECIAL MEETING OF FOREIGN AND DEFENSE MINISTERS OF NATO: THEATER NUCLEAR FORCES [TNF], DECEMBER 12, 1979 (EXCERPT)[1]

. . .6. Ministers noted that these recent developments require concrete actions on the part of the Alliance if NATO's strategy of flexible response is to remain credible. After intensive considerations, including the merits of alternative approaches, and after taking note of the positions of certain members, Ministers concluded that the overall interest of the Alliance would best be served by pursuing two parallel and complementary approaches of TNF modernization and arms control.

7. Accordingly Ministers have decided to modernize NATO's LRTNF by the deployment in Europe of US ground-launched systems comprising 108 Pershing II launchers, which would replace existing US Pershing I-A, and 464 GLCM [ground-launched cruise missiles],[2]

[1]United States Arms Control and Disarmament Agency, *Documents on Disarmament, 1979* (Washington, D.C.: U.S. Government Printing Office, 1982), pp. 814–16.

[2]Brackets throughout this document are in original.

all with single warheads. All the nations currently participating in the integrated defense structure will participate in the program: the missiles will be stationed in selected countries, and certain support costs will be met through NATO's existing common funding arrangements. The program will not increase NATO's reliance upon nuclear weapons. In this connection, Ministers agreed that as an integral part of TNF modernization, 1,000 US nuclear warheads will be withdrawn from Europe as soon as feasible. Further, Ministers decided that the 572 LRTNF warheads should be accommodated within that reduced level, which necessarily implies a numerical shift of emphasis away from warheads for delivery systems of other types and shorter ranges. In addition they noted with satisfaction that the Nuclear Planning Group [NPG] is undertaking an examination of the precise nature, scope, and basis of the adjustments resulting from the LRTNF deployment and their possible implications for the balance of roles and systems in NATO's nuclear armor as a whole. This examination will form the basis of a substantive report to NPG Ministers in the Autumn of 1980.

8. Ministers attach great importance to the role of arms control in contributing to a more stable military relationship between East and West and in advancing the process of détente. This is reflected in a broad set of initiatives being examined within the Alliance to further the course of arms control and detente in the 1980s. They regard arms control as an integral part of the Alliance's efforts to assure the undiminished security of its member states and to make the strategic situation between East and West more stable, more predictable, and more manageable at lower levels of armaments on both sides. In this regard they welcome the contribution which the SALT II treaty makes towards achieving these objectives.

9. Ministers consider that, building on this accomplishment and taking account of the expansion of Soviet LRTNF capabilities of concern to NATO, arms control efforts to achieve a more stable overall nuclear balance at lower levels of nuclear weapons on both sides should therefore now include certain US and Soviet long-range theater nuclear systems. This would reflect previous Western suggestions to include such Soviet and US systems in arms control negotiations and more recent expressions by Soviet President Brezhnev of willingness to do so. Ministers fully support the decision taken by the United States following consultations within the Alliance to negotiate arms limitations on LRTNF and to propose to the USSR to begin negotiations as soon as possible along the following lines which have been elaborated in intensive consultations within the Alliance:

A. Any future limitations on US systems principally designed for theater missions should be accompanied by appropriate limitations on Soviet theater systems.

B. Limitations on US and Soviet long-range theater nuclear systems should be negotiated bilaterally in the SALT III framework in a step-by-step approach.

C. The immediate objective of these negotiations should be the establishment of agreed limitations on US and Soviet land-based long-range theater nuclear missile systems.

D. Any agreed limitations on these systems must be consistent with the principle of equality between the sides. Therefore, the limitations should take the form of de jure equality both in ceilings and in rights.

E. Any agreed limitations must be adequately verifiable.

10. Given the special importance of these negotiations for the overall security of the Alliance, a special consultative body at a high level will be constituted within the Alliance to support the US negotiating effort. This body will follow the negotiations on a continuous basis and report to the Foreign and Defense Ministers who will examine developments in these negotiations at their semi-annual meetings.

11. The Ministers have decided to pursue these two parallel and complementary approaches in order to avert an arms race in Europe caused by the Soviet TNF build-up, yet preserve the viability of NATO's strategy of deterrence and defense and thus maintain the

security of its member states. A modernization decision, including a commitment to deployments, is necessary to meet NATO's deterrence and defense needs, to provide a credible response to unilateral Soviet TNF deployments, and to provide the foundation for the pursuit of serious negotiations on TNF. Success of arms control in constraining the Soviet build-up can enhance Alliance security, modify the scale of NATO's TNF modernization requirements, and promote stability and détente in Europe in consonance with NATO's basic policy of deterrence, defense and détente as enunciated in the Harmel Report, NATO's TNF requirements will be examined in the light of concrete results reached through negotiations.

PRESIDENT RONALD REAGAN'S SPEECH TO THE UNITED NATIONS GENERAL ASSEMBLY, September 26, 1983 (EXCERPTS)[1]

. . .We are determined to spare no effort to achieve a sound, equitable and verifiable agreement. For this reason, I have given new instructions to Ambassador Nitze in Geneva, telling him to put forward a package of steps designed to advance the negotiations as rapidly as possible. These initiatives build on the interim framework the United States advanced last March and address concerns that the Soviets have raised at the bargaining table in the past. Specifically:

- First, the United States proposes a new initiative on global limits. If the Soviet Union agrees to reductions and limits on a global basis, the United States for its part, will not offset the entire Soviet global missile deployment through US deployments in Europe. We would, of course, retain the right to deploy missiles elsewhere.
- Second, the United States is prepared to be more flexible on the content of the current talks. The United States will consider mutually acceptable ways to address the Soviet desire that an agreement should limit aircraft as well as missiles.
- Third, the United States will address the mix of missiles that would result from reductions. In the context of reductions to equal levels, we are prepared to reduce the number of *Pershing* II ballistic missiles as well as ground-launched cruise missiles.

I have decided to put forward these important initiatives after full and extensive consultations with our allies, including personal correspondence I have had with the leaders of the NATO Governments and Japan and frequent meetings of the NATO Special Consultative Group. I have also stayed in close touch with other concerned friends and allies. The door to an agreement is open. It is time for the Soviet Union to walk through it.

I want to make an unequivocal pledge to those gathered today in this world arena. The United States seeks and will accept any equitable, verifiable agreement that stabilizes forces at lower levels than currently exist. We are ready to be flexible in our approach, indeed, willing to compromise. We cannot, however, especially in light of recent events, compromise on the necessity of effective verification.

Reactions to the Korean airliner tragedy are a timely reminder of just how different the Soviets' concept of truth and international co-operation is from that of the rest of the world. Evidence abounds that we cannot simply assume that agreements negotiated with the Soviet Union will be fulfilled. . . .

Peace cannot be served by pseudo arms control. We need reliable, reciprocal reductions.

---

[1]*Survival* (January–February 1984), pp. 32–33.

I call upon the Soviet Union today to reduce the tensions it has heaped on the world in the past few weeks, and to show a firm commitment to peace by coming to the bargaining table with a new understanding of its obligations. I urge it to match our flexibility.

If the Soviets sit down at the bargaining table seeking genuine arms reductions, there will be arms reductions. The governments of the West and their people will not be diverted by misinformation and threats. The time has come for the Soviet Union to show proof that it wants arms control in reality, not just in rhetoric. . . .

NATO NUCLEAR PLANNING GROUP: COMMUNIQUÉ AND ANNEX [THE MONTEBELLO DECISION], October 28, 1983 (EXCERPTS)[1]

. . . Ministers noted that, in contrast to NATO's policy of restraint, the Soviet build-up is continuing relentlessly at all levels. In recent months the Soviet Union has continued construction of at least three new SS-20 bases east of the Ural Mountains, in addition to the already deployed 351 operational SS-20 launchers comprising 1,053 warheads. They are also retaining a large number of SS-4 and SS-5 missiles and are replacing older shorter-range missiles by a new generation of more accurate systems, namely the SS-21, SS-22 and SS-23, which have been under development for a number of years. SS-21s have already been forward deployed in Eastern Europe for some time. Ministers noted that the Soviet Union has recently threatened to deploy forward additional shorter-range missiles and to adopt other military and political measures should NATO proceed as planned with the deployment of LRINF in Europe. Ministers urged the Soviet Union to match NATO's restraint and to reduce rather than increase the levels of its nuclear weapons. . . .

Ministers emphasized NATO's commitment and determination that, in the absence of a concrete arms-control agreement obviating the need for that deployment, deployment of a mixed force of ballistic and cruise missiles as decided in the dual-track decision of 1979 will begin by the end of the year. . . .

Ministers welcomed the latest United States initiatives which address concerns that the Soviet Union has expressed in Geneva on aircraft, geographic scope and reductions of the *Pershing* II as well as cruise missiles. The Soviet Union has not yet demonstrated similar flexibility. Each Soviet modification of their original proposal made so far would have the same basic outcome: the prohibition of any NATO LRINF missile deployments while the Soviet Union retained a monopoly in such missiles. This is because of their insistence to take into account British and French forces and their refusal to accept the principle of equality in LRINF missile deployments. The Alliance cannot accept this Soviet approach which is apparently intended to further their long-term aim of dividing the Alliance and to erode seriously the linkage between the United States strategic deterrent and the defence of NATO Europe. This linkage is essential for NATO's security.

Ministers noted with attention the most recent Soviet proposals. The Soviets are invited to explain fully their proposals at the negotiating table. Ministers stated that NATO would continue to examine carefully all proposals in light of the Alliance's agreed criteria described above. As reported, however, the Soviet Union regrettably continues to insist on a monopoly in land-based LRINF missiles. The West has continued to negotiate for two years while the Soviet SS-20 build-up has continued. The Soviets therefore have absolutely no justification for breaking off negotiations as NATO's deployment goes forward, as they have threatened.

---

[1]*Survival* (January–February 1984), pp. 36–37.

Ministers stressed that the Alliance commitment to negotiations would continue even after initial deployments and that they remain willing to reverse, halt or modify deployments—including the removal and dismantling of missiles already deployed—if and when a balanced, equitable and verifiable agreement is achieved in Geneva. . . .

The Danish Minister of Defence repeated Danish Government support for the double-track decision but at the same time he recalled the motion passed by the Folketing on May 26.

Greece has expressed its views in a statement in the minutes.

Annex: The Montebello Decision

. . .the Alliance since 1977 has been conducting analyses aimed at assuring that nuclear weapons in NATO's armoury are held to the minimum number necessary for deterrence, taking account of developments in conventional as well as nuclear forces. . . .

With the Alliance analysis now complete, the Nuclear Planning Group has decided on 27 October 1983 to withdraw 1,400 warheads during the next several years. This Ministerial decision, taken together with the already accomplished withdrawal of 1,000 warheads will bring to 2,400 the total number of warheads to be removed from Europe since 1979. Moreover, this reduction will not be affected by any deployment of Longer-Range INF (LRINF) since one warhead will be removed for each *Pershing* II or Ground-Launched Cruise Missile (GLCM) warhead deployed.

The detailed implementation of this decision as to the precise composition of the stockpile is a matter for the responsible military authorities to determine and a programme to effect this will be worked out and implemented over the next five to six years. In this context, appropriate consideration will be given to short-range systems. NATO's military authorities should report their findings at a future NPG meeting.

Recognizing that for this minimum level stockpile to make the most effective contribution to deterrence, both the delivery systems and the warheads must be survivable, responsive and effective, Ministers accordingly identified a range of possible improvements. Ministers established broad criteria which will remain valid for the next decade, including the continuing importance of strengthening conventional forces. The Alliance must, however, take account at all times of changes to Soviet capabilities during this period.

Contrary to the impression that NATO has been fuelling an arms build-up by adding to its nuclear armoury, this sustained programme of reductions will have reduced NATO's nuclear stockpile to the lowest level in over 20 years. Ministers urged the Soviet Union to follow the example set by the Alliance, to halt and reverse its build-up of nuclear forces, and to join NATO in the search for a safer future.

SPEECH BY FRG CHANCELLOR HELMUT SCHMIDT TO THE SPD CONFERENCE, November 19–20, 1983 (EXCERPT)[1]

. . .While recognizing the failures of both super-powers, I still believe firmly in the dual-track decision of December 1979—not because I was its supposed "inventor", not only because I myself worked on the indissoluble connexion between the elements of deployment and arms talks, but because I still see it as an effective instrument for the restoration of balance through the limitation and step-by-step removal of the nuclear weapons in Europe.

---

[1]*Survival* (March–April 1984), pp. 87–88.

Two conversations with Leonid Brezhnev deeply impressed me: he spoke of the suffering of the last war, of the sacrifices of his people and those of his own family. His wish for peace was clear to me. At the same time I have not forgotten the Brezhnev doctrine. And since then, military thinking has had more influence on the Soviet leadership.

The Helsinki Agreement in 1975 was up till now the high point of the double strategy of deterrence plus co-operation with the Soviet Union and détente. The concept of "containment" of the late 1940s had at last been realized, but had, however, been added to in an important way. The West accepted in Helsinki that coexistence of states would in no way mean ideological coexistence—in other words, the Russians were not offering a halt to Soviet infiltration all over the world.

I can understand the worries of the Soviet military leadership about the Germans, Americans and Chinese, but I cannot in any way accept them as reasonable. These three compounded threats however do not suffice to explain their armament; that is explained also by Russia's inherent expansionist tendency. The Andropov leadership basically does not differ in this.

The Russians do not want to understand that their persecution complex, their security complex, their striving for absolute security leads to ever more insecurity for Europeans, for Americans, Chinese, Japanese and so on. They also do not understand that Europeans, Americans, Chinese and Japanese can in no way accept this growing insecurity. The Russians have seen that the West does not always react as one to threats. They hope to be able to separate the West European allies and the US psychologically and politically, or even to play them off against each other. Naturally, we Germans are one of the main targets of Soviet propaganda.

The exaggerated build-up of Soviet SS-20s was a grave mistake. The rejection of the "walk-in-the-woods" formula by the US and Soviet negotiators Nitze and Kvitsinsky was a second mistake. But the walk in the woods, an acceptable compromise between two irreconcilable negotiating positions, was also accompanied by mistakes by the Americans. First, they turned down this compromise. Secondly, they did this without informing their European allies and the countries that were accepting missiles, and failing then to consult them. For instance, in the last ten weeks until my leaving office I learnt nothing of this. There is no excuse for the failure to inform and consult the European allies. It is a grave violation of the agreed procedure of the dual-track decision.

I must say that this mistake disturbs me, as do also ways of speaking about the pre-war phase or the possibility of conducting a limited nuclear war. Some speeches of US politicians on the conservative wing of the Republican Party which were meant to give the Soviet leadership cause for thought have thrown Europeans into uncertainty about what the strategic aims of the Administration actually are.

But for me one thing is clear: so long as Russian forces stand in the eastern part of Central Europe, the US must also play a role in Western Europe.

This is now my first reason for deployment: the Federal Republic must keep its word—despite all the disappointments over speeches and behavior in Washington. My second reason is that the political balance would be permanently upset if the Soviet Union succeeded with its unprovoked arms build-up. I found widespread agreement when I said on 1 October last year—my last words in Parliament as Chancellor—"everyone should and must be able to count on our constancy". I have not changed my opinion.

It remains clear to me that only as many Western systems should be deployed as it takes for Moscow to know that:

- The Federal Republic would fulfil its duties to the Alliance.

- We are ready to give the US the rights in Western Europe necessary for common security.
- During any attempt to put Europe under pressure, the Soviet Union must assume that such threats will be taken as directed to the American systems in Western Europe, so that the US will therefore be automatically involved.

For me, it was also and remains clear that there should never be so many US systems stationed in Western Europe that anyone in Washington could seriously come to think that a nuclear confrontation with the Soviet Union could be limited to Europe.

We Social Democrats should not isolate ourselves in the opposition. We should in particular take into consideration the views of governments in London and Paris. If anyone in Europe can understand and satisfy our special German concerns then it is them. . . .

STATEMENT BY GENERAL SECRETARY ANDROPOV, November 24, 1983 (EXCERPTS)[1]

. . .Washington, Bonn, London and Rome did not heed the voice of reason—the deployment of American medium-range missiles is beginning on the territory of the FRG, Britain and Italy. Thereby the appearance of American *Pershings* and cruise missiles in the European continent is becoming an accomplished fact. . . .

Can the Soviet Union and the other Socialist countries ignore this danger? No, they cannot. That is why the highest party and state leaders of seven Socialist countries declared at their meeting in Moscow on 28 June 1983 that under no circumstances will they permit a military superiority of the NATO bloc over the countries of the Warsaw Treaty.

When confirming their consent to the deployment of American missiles in their countries the governments of the FRG, Britain and Italy could not but know that it was from the outset that the United States did not want the attainment of a mutually acceptable accord on nuclear arms in Europe and did everything at the talks in Geneva and outside them to prevent such an accord. Likewise they could not but know that the Soviet Union [and] its allies would definitely take the necessary measures to safeguard their security, to prevent the United States and NATO as a whole from breaking the existing rough balance of forces in Europe.

It was also clearly stated by us that the appearance of new American missiles in Western Europe would make impossible a continuation of the talks that were being conducted in Geneva on nuclear arms in Europe.

The decisions adopted during the past few days by the governments of the FRG, Britain and Italy unequivocally show that, contrary to the will of their own peoples, contrary to the interests of the security of their countries and contrary to the interests of European and universal peace, these governments have given the green light to the deployment of the American missiles. Thereby they have assumed, together with the government of the United States, the entire responsibility for the consequences of the myopic policy about which the Soviet Union had warned in advance.

Having thoroughly weighed all the aspects of the obtaining situation the Soviet leadership adopted the following decisions:

First, since by its actions the United States has torpedoed the possibility of reaching a

---

[1] *Survival* (March–April 1984), pp. 89–90.

mutually acceptable accord at the talks on questions of limiting nuclear arms in Europe, and their continuation in these conditions would only serve as a cover for the actions of the United States and . . . other NATO countries directed at undermining European and international security, the Soviet Union considers its further participation in these talks impossible.

Second, being cancelled are the Soviet Union's unilateral obligations which had for their purpose the creation of more favourable conditions for success at the talks. Thereby the moratorium on the deployment of Soviet medium-range nuclear weapons in the European part of the USSR is abrogated.

Third, on agreement with the governments of the GDR and Czechoslovakia, the announced preparatory work to deploy on the territory of these countries operational–tactical missiles of increased range that was started some time ago will be accelerated.

Fourth, since by deploying its missiles in Europe the United States increases the nuclear threat to the Soviet Union, the corresponding Soviet systems will be deployed with due account for this circumstance in ocean areas and in seas. By their characteristics these systems of ours will be adequate to the threat which is being created to us and our allies by the American missiles that are being deployed in Europe.

It goes without saying that other measures, too, will be taken to ensure the security of the USSR and the other countries of the Socialist community.

As we commence the fulfilment of the decisions adopted by us, we state that the reply measures from the Soviet side will be kept strictly within the limits that will be dictated by the actions of the NATO countries. The Soviet Union, and we stress this again, does not strive for military superiority, and we will do only what is absolutely necessary to prevent the military balance from being disrupted.

Should the United States and the other NATO countries display readiness to return to the situation that existed before the commencement of the deployment of the American medium-range missiles in Europe, the Soviet Union will also be prepared to do this. In that event our earlier proposals on questions of limiting and reducing nuclear arms in Europe would become valid again. In that case—that is, on condition of the restoration of the former situation—the USSR's unilateral obligations in this field would also become effective again. . . .

The Soviet leadership calls on the leaders of the United States and the West European countries to weigh once again all the consequences with which their own peoples and the whole of mankind are being threatened by the fulfilment of the plans to deploy the new American missiles in Europe. . . .

COMMUNIQUÉ OF THE NORTH ATLANTIC COUNCIL, June 11, 1987 (EXCERPTS)[1]

. . . Having reviewed progress in the negotiations between the United States and the Soviet Union on an INF agreement, the Allies concerned call on the Soviet Union to drop its demand to retain a portion of its SS-20 capability and reiterate their wish to see all long-range land-based missiles eliminated in accordance with NATO's long-standing objective.

They support the global and effectively verifiable elimination of all US and Soviet land-based SRINF missiles with a range between 500 and 1,000 km as an integral part of an INF agreement.

They consider that an INF agreement on this basis would be an important element in a

[1]*Survival* (September–October 1987), pp. 461–62.

coherent and comprehensive concept of arms control and disarmament which, while consistent with NATO's doctrine of flexible response, would include:

- a 50% reduction in the strategic offensive nuclear weapons of the US and the Soviet Union, to be achieved during current Geneva negotiations;
- the global elimination of chemical weapons;
- the establishment of a stable and secure level of conventional forces, by the elimination of disparities, in the whole of Europe;
- in conjunction with the establishment of a conventional balance and the global elimination of chemical weapons, tangible and verifiable reductions of American and Soviet land-based nuclear missile systems of shorter range, leading to equal ceilings.

We have directed the North Atlantic Council in Permanent Session, working in conjunction with the appropriate military authorities, to consider the further development of a comprehensive concept of arms control and disarmament. The arms-control problems faced by the Alliance raise complex and interrelated issues which must be evaluated together, bearing in mind overall progress in the arms-control negotiations enumerated above as well as the requirements of Alliance security and of its strategy of deterrence. . . .

NATO NUCLEAR PLANNING GROUP COMMUNIQUÉ, NOVEMBER 1987[1]

1. The NATO Nuclear Planning Group (NPG) met in Ministerial session in Monterey, California, on 3rd and 4th November, 1987. Iceland attended as an observer. We discussed a variety of security matters pertaining to NATO's nuclear forces, such as current arms control negotiations, the status of implementation of the December 1979 Dual-Track and 1983 Montebello Decisions, the work of several study groups and future NPG work.

2. The fundamental security objectives of the Alliance are to deter aggression and to provide an environment for peaceful and stable relationships between East and West on the basis of balanced forces at the lowest possible level. The maintenance of effective military forces and the pursuit of arms control are complementary elements of this security policy.

3. We welcome and fully support the agreement in principle between the United States and the Soviet Union for the global elimination of land-based INF missiles with a range between 500 and 5,500 kms. This has been made possible by the determination and solidarity of the Allied Governments over the years. We look forward to the prospect of a verifiable INF treaty being signed and ratified in the near future.

4. With the prospect of Soviet agreement to long standing Alliance INF arms control objectives, we now look forward to rapid progress in the START negotiations and re-affirm our support for the 50% reduction in the strategic nuclear arsenals of the United States and the Soviet Union as proposed by the United States. In this connection, we emphasize the importance of a comprehensive, integrated and coherent approach to all elements of arms control and security, nuclear and non-nuclear.

5. Our strategy of flexible response will continue to be vital to the security of the Alliance. We remain concerned about the offensive capabilities of the Warsaw Pact arrayed against us. We are therefore determined, consistent with the framework of the Montebello

[1]*NATO Review* (December 1987), p. 27.

Decision and with our arms control obligations, to continue to implement those measures required to maintain the effectiveness, responsiveness and survivability of our nuclear forces. In doing so, it is and will remain Alliance policy to possess only the minimum number of nuclear weapons necessary for a credible deterrent. In accordance with this policy, and in spite of the fact that Soviet nuclear forces have continued to increase during that period, the Alliance has already unilaterally reduced its nuclear stockpile in Europe to the lowest level in over twenty years.

6. We accepted with pleasure the invitation of the Danish Government to hold our next Nuclear Planning Group Ministerial meeting in Denmark in the Spring of 1988.

7. Greece has expressed its views in a statement included in the Minutes.

NORTH ATLANTIC COUNCIL MINISTERIAL COMMUNIQUÉ, DECEMBER 1987[1]

The North Atlantic Council met in Ministerial Session in Brussels on 11th December 1987. Ministers heard, and exchanged views on, a detailed report by the United States Secretary of State on the Washington summit talks between President Reagan and General Secretary Gorbachev, including the conclusion of a Treaty without precedent in the history of arms control which will eliminate all land-based intermediate-range nuclear missiles of the two parties, and notable progress in other areas.

Ministers agreed as follows:

1. We welcome the Washington INF Treaty. This Treaty is the successful result of solidarity and resolve shown by the Alliance as a whole. It is fully consistent with the security requirements of the Alliance. It accomplishes an important and long-standing Alliance objective: the elimination of a class of Soviet nuclear weapons threatening the European Allies and other regions of the world. The Treaty establishes significant new standards in verification procedures and in achieving asymmetric reductions to redress the present imbalance. The process of intense consultation and co-operation within the Alliance in support of this effort has demonstrated and reinforced the close ties and common interests between Europe and North America that lie at the heart of the North Atlantic Alliance. We consider that the INF Treaty will be all the more meaningful because it opens the way to progress in other arms control areas. We fully support the INF Treaty and urge its early entry into force.

2. We are convinced that these developments have been made possible by the consistent and realistic pursuit of Alliance security policy set forth in the Harmel Report whose 20th Anniversary we celebrate this year. We are agreed upon the importance of Alliance cohesion and solidarity, and adequate military strength, which remain an essential basis for our policy of dialogue and co-operation aimed, together with arms control, at resolving the underlying tensions between East and West. We take this opportunity to reaffirm the continuing validity of this balanced and complementary approach, which has ensured Alliance security, prevented war and helped bring about a constructive East–West dialogue.

Notwithstanding the continuing fundamental differences between the states of East and West, we note a more forthcoming Soviet attitude, which could lead to real progress in East–West relations, especially in the field of arms control. We are ready to explore all possibilities thus offered to us with realism and open-mindedness. We shall do so, thoroughly investigating areas of common interest and acting on the basis of actual Soviet and Warsaw

[1]*NATO Review* (December 1987), pp. 24–25.

Pact conduct and of a sober calculation of the implications for our own security. Only in this way can the unnatural division of Europe be healed. As the Harmel Report states, the ultimate political purpose of our Alliance is to achieve a just and lasting peaceful order in Europe accompanied by appropriate security guarantees.

3. An essential element of the Alliance is our commitment to share both the burden and the benefits of our common effort. A sustained and adequate contribution by all concerned is imperative to our collective security. We agree on the undiminished need to respond in solidarity to a continuing Soviet military threat, sustained by current Soviet armaments programmes.

For the foreseeable future, there is no alternative to the Alliance concept for the prevention of war, which is a strategy of deterrence based upon an appropriate mix of adequate and effective nuclear and conventional forces, each element being indispensable. It is on this basis that the countries of the Alliance will be able to continue their own essential contribution to peace and security and to deterring war and threat of aggression or intimidation.

4. Arms control is an integral part of this security policy. We will continue to seek effectively verifiable arms control measures leading to a more stable and secure balance of forces at lower levels. In light of the decisions they took at the Reykjavik Ministerial of last June, the fifteen Allies concerned consider that the INF Treaty is an important element in a coherent and comprehensive concept of arms control and disarmament, consistent with NATO's doctrine of flexible response, which would include:

- a 50% reduction in the strategic offensive nuclear weapons of the US and Soviet Union;
- the global elimination of chemical weapons;
- the establishment of a stable and secure level of conventional forces, by the elimination of disparities, in the whole of Europe;
- in conjunction with the establishment of a conventional balance and the global elimination of chemical weapons, tangible and verifiable reductions of American and Soviet land-based nuclear missile systems of shorter-range, leading to equal ceilings.

5. We further note that the North Atlantic Council in Permanent Session is actively continuing its work, in conjunction with the appropriate military authorities, to consider the further development of a comprehensive concept of arms control and disarmament. We recognise that the arms control problems faced by the Alliance raise complex and interrelated issues which must be evaluated together, bearing in mind overall progress in the arms control negotiations enumerated above as well as the requirements of Alliance security and of its strategy of deterrence.

6. We look to fresh progress with a view to the speedy conclusion of an agreement in the START context, in line with the Reykjavik goal of achieving a 50% reduction in the Soviet and United States strategic arsenals.

7. We hope that the recent Soviet recognition of the need for comprehensive verification of a chemical weapons ban can soon be translated into treaty obligations providing for effective verification, and that a convention on the global elimination of these weapons and their future non-production can be quickly concluded. We reaffirm our commitment to two future security negotiations, in the framework of the CSCE process, on conventional stability aiming at a comprehensive, stable and verifiable balance of conventional forces at lower levels and new confidence building and security measures in Europe, from the Atlantic to the Urals. We hope that agreements on mandates for such negotiations meeting our objectives can be achieved as part of a balanced outcome of the Vienna CSCE follow-on meeting so that the new negotiations can begin in the near future.

8. Those of us participating in the MBFR [Mutual and Balanced Force Reductions] negotiations reiterate their desire to reach an early, substantial and verifiable agreement and call upon the Warsaw Pact participants to adopt a more constructive posture in the negotiations.

9. Progress towards more security and peace requires not only specific measures in the field of arms control, but also substantial changes in the overall East–West relationship, in line with the principles of the Helsinki Final Act. The Allies thus strongly support the continuation and strengthening of the CSCE process as a vital instrument to develop and enhance stable and constructive East–West relations on a long-term basis. We are concerned about the slow progress at the current CSCE follow-up meeting and the essential need for the USSR and its allies to improve significantly their implementation of the CSCE Documents, especially in the human dimension. The results of the Vienna follow-up meeting must not be confined only to the military aspects of security; it is essential for balanced progress to be recorded in all spheres covered by the Helsinki Final Act. We recall the proposals to this effect submitted by Western allies. In this respect we attach particular importance to the Western proposal on the human dimension. We will continue to work for the speedy adoption of a substantive concluding document containing concrete improvements with a view to further implementing the Final Act as well as the Madrid Concluding Document and assuring a proper balance between all CSCE areas.

We note with satisfaction the implementation up to now of the Stockholm Document.

10. We note with satisfaction that the Foreign and Defence Ministers of the Western European Union in their Declaration published at The Hague on 27th October 1987 underlined a number of our basic principles and that they affirmed a positive identity in the field of European security within the framework of the Atlantic Alliance, conducive to the strengthening of the transatlantic partnership and of the Alliance as a whole.

11. The maintenance of a calm situation in and around Berlin, including unhindered traffic on all access routes, remains of fundamental importance for East–West relations. The viability and attractiveness of Berlin were impressively underlined during the 750th anniversary year. The members of the Alliance welcome current efforts to improve and expand Berlin's air services, to enhance the city's role as a centre for international events, and to strengthen human contacts there.

The Alliance supports the efforts of the Federal Republic of Germany to promote a permanent dialogue and co-operation with the GDR, for the benefit of the divided German people, including Berlin within this process. These efforts contribute to furthering East–West co-operation and strengthening peace and stability in Europe.

12. We deplore the Soviet occupation of Afghanistan and call, in accordance with the United Nations repeated resolutions on this subject, for a speedy and complete withdrawal of Soviet troops and the installation of a transitional government acceptable to all parties.

13. We approved the establishment of a NATO Conventional Armaments Planning System on a trial basis, subject to the completion of the necessary supporting policy guidance. The new system will encourage nations to get better value out of the resources they devote to defence procurement and thereby better to support the Alliance by providing a useful framework within which it will be possible to analyse long term national equipment requirements, to examine to what extent national plans meet Alliance needs and to promote greater co-operation in defence equipment.

We shall continue to pursue actively our efforts to enhance armaments co-operation among all the countries of the Alliance. The work of the IEPG [Independent European Programme Group] represents an important contribution to the transatlantic dialogue in this field.

14. In the spirit of Article 2 of the North Atlantic Treaty and in view of the prosperity gap among the various members of the Alliance, we reaffirm the importance of improving the present level of mutual assistance and co-operation within the Alliance.

15. We condemn terrorism in all its forms. We reiterate our resolve to fight against it and reaffirm our belief that closer international co-operation remains central to defeating this scourge.

16. Ministers noted with regret the intention of the Secretary General, Lord Carrington, to relinquish his post. They invited Mr. Manfred Wörner to become Secretary General of the Organization as from 1st July 1988 and expressed satisfaction at Mr. Wörner's acceptance.

17. The Spring 1988 meeting of the Council in Ministerial Session will be held in Madrid on 9th and 10th June.

Arms Control: Short-Range Nuclear Missiles

STATEMENT BY BRITISH FOREIGN AND COMMONWEALTH OFFICE, May 14, 1987 (EXCERPTS)[1]

In answer to a question about the British position on an INF agreement, Spokesman said that an agreement on LRINF must include constraints on shorter-range systems. This is a long-standing NATO requirement. Spokesman added that this could be met by equal ceilings at various levels down to zero; the British Government could accept zero/zero SRINF provided that the conditions adequately safeguard Western security. NATO is now studying what level best meets the requirements of the Alliance as a whole; NATO's final position will be settled in these consultations.

NATO NUCLEAR PLANNING GROUP COMMUNIQUÉ, May 15, 1987 (EXCERPTS)[2]

. . .Deterrence of any aggression continues to be the central security objective of the Alliance. To that end, in this the year of the twentieth anniversary of the adoption of the strategy of flexible response, we noted that this strategy has stood the test of time and remains an essential and sound basis for the future security of all Alliance members. While improving NATO's conventional forces, we will maintain and improve the nuclear forces necessary to carry out that strategy. In that context, we noted with concern the existing imbalances between Alliance and Warsaw Pact nuclear, conventional and chemical forces as well as the unabated expansion of Warsaw Pact military capabilities across the board.

Efforts to secure equitable and effectively verifiable reductions in military forces, both conventional and nuclear, are an integral element of our security policy in seeking to achieve a more stable and secure environment at lower levels of armaments. It is in our security interests that agreements ensure detailed, specific arrangements providing for effective ver-

---

[1]*Survival* (September–October 1987), p. 459.

[2]*Survival* (September–October 1987), p. 458.

ification; we reject generalized undertakings on verification as an acceptable basis for sound agreements.

During our continuing consultations on INF arms control, we welcomed the improved prospects for a longer-range INF agreement between the United States and the Soviet Union encompassing significant reductions in nuclear forces. We reaffirmed that appropriate global constraints on shorter-range missile systems are indispensable. We stressed the requirement to eliminate all United States and Soviet LRINF missiles and called upon the Soviet Union to drop its demand to retain a portion of its SS-20 force. A global zero outcome, a long-standing NATO objective, would further reduce the Soviet threat, and greatly facilitate verification.

STATEMENT BY THE GOVERNMENT OF THE FEDERAL REPUBLIC OF GERMANY, June 2, 1987[1]

Our paramount political objective is to safeguard peace and prevent any war, be it conventional or nuclear. It is this consideration that prompted the Alliance to develop its strategy of flexible response, which has proved its worth. It is a strategy aimed at preventing war—it has done so until now and will continue to do so. There is no alternative to this strategy in the foreseeable future. Its effectiveness and credibility are dependent on balanced nuclear and conventional forces. For this reason, the Alliance cannot completely dispense with nuclear weapons in the foreseeable future. Equally indispensable to our strategy of preventing war is the presence of allied troops in Europe, especially US forces.

In the consistent pursuit of its policy for peace, the Federal Government advocates that a stable military balance in Europe affording security for all countries be established by means of extensive disarmament in all areas.

We strive for arms-control agreements which guarantee increased security for everyone concerned at the lowest possible balanced level of forces. This requires that the legitimate security needs of everybody affected—large as well as medium and small countries—be taken into account. Our Alliance is based on the principle that one country's security is simultaneously the other countries' security. There must be no forms of lesser security either in the relationship between the American and European allies or in relations among the European allies.

In keeping with this policy and pursuant to the Alliance's decisions of 1979 and 1983 (Montebello), a total of 2,400 US nuclear warheads are being withdrawn, unilaterally and without replacement, from Europe in the period from 1980 to 1988. This will reduce NATO's nuclear arsenal to the lowest level in over 20 years.

The Federal Government supports the swift elimination of all land-based INF missiles with longer ranges between 1,000 and 5,000 km (LRINF systems) by the United States and the Soviet Union, ideally on a global scale. This would be the first far-reaching step towards disarmament. This achievement would above all be the outcome of the firm and united position taken by the Federal Government and the entire Alliance in implementing the NATO two-track decision. It would simultaneously be an indication that tangible steps towards disarmament are attainable on the basis of an equitable reconciliation of the two sides' interests.

The Federal Government advocates that the US and Soviet land-based nuclear systems with ranges up to 1,000 km still left in Europe be tangibly and verifiably reduced through

[1]*Survival* (September–October 1987), pp. 459–60.

negotiations on a step-by-step basis in conjunction with the establishment of conventional equilibrium and the world-wide elimination of chemical weapons.

An initial step would be an agreement between the United States and the Soviet Union with global validity to the effect that in future they will not possess any INF missiles with shorter ranges between 500 and 1,000 km (SRINF systems). The Federal Government urges that success along these lines be achieved as quickly as possible in the US–Soviet negotiations at Geneva.

The Federal Government maintains its position that the 72 *Pershing* IA missiles of the *Bundeswehr* are not and cannot be covered by the ongoing negotiations. The Federal Government is at one on this point with the United States and its other allies.

The Federal Government considers the following to be necessary and indeed feasible as the next steps along the path towards far-reaching disarmament:

- 50% reduction of the strategic nuclear arsenals of the Soviet Union and the United States.
- Global elimination of chemical weapons.
- Reduction of conventional land and air forces in Europe from the Atlantic to the Urals to a balanced low level. This presupposes the elimination of the Warsaw Pact's conventional superiority.
- Further substantive cuts in the nuclear arsenals of the United States and the Soviet Union remaining in Europe.

Arms Control: Conventional Forces

DOCUMENT OF THE STOCKHOLM CONFERENCE ON CONFIDENCE- AND SECURITY-BUILDING MEASURES AND DISARMAMENT IN EUROPE, September 19, 1986 (EXCERPTS)[1]

. . . Prior Notification of Certain Military Activities

The participating States will give notification in writing through diplomatic channels in an agreed form of content, to all other participating States 42 days or more in advance of the start of notifiable military activities in the zone of application for confidence- and security-building measures (CSBM).

Notification will be given by the participating State on whose territory the activity in question is planned to take place even if the forces of that State are not engaged in the activity or their strength is below the notifiable level. This will not relieve other participating States of their obligation to give notification, if their involvement in the planned military activity reaches the notifiable level.

Each of the following military activities in the field conducted as a single activity in the zone of application for CSBM at or above the levels defined below, will be notified:

> The engagement of formations of land forces [amphibious, air mobile and airborne] of the participating States in the same exercise activity conducted under a single operational command independently or in combination with any possible air or naval components.

This military activity will be subject to notification whenever it involves at any time during the activity:

- at least 13,000 troops, including support troops, or
- at least 300 battle tanks

if organised into a divisional structure or at least two brigades/regiments, not necessarily subordinate to the same division.

---

[1] *Survival* (January–February 1987), pp. 79–84.

The participation of air forces of the participating States will be included in the notification if it is foreseen that in the course of the activity 200 or more sorties by aircraft, excluding helicopters, will be flown.

The engagement of military forces either in an amphibious landing or in a parachute assault by airborne forces in the zone of application for CSBM.

These military activities will be subject to notification whenever the amphibious landing involves at least 3,000 troops or whenever the parachute drop involves at least 3,000 troops.

The engagement of formations of land forces of the participating States in a transfer from outside the zone of application for CSBM to arrival points in the zone, or from inside the zone of application for CSBM to points of concentration in the zone, to participate in a notifiable exercise activity or to be concentrated.

The arrival or concentration of these forces will be subject to notification whenever it involves, at any time during the activity:

- at least 13,000 troops, including support troops, or
- at least 300 battle tanks

if organised into a divisional structure or at least two brigades/regiments, not necessarily subordinate to the same division.

Forces which have been transferred into the zone will be subject to all provisions of agreed CSBM when they depart their arrival points to participate in a notifiable exercise activity or to be concentrated within the zone of application for CSBM.

Notifiable military activities carried out without advance notice to the troops involved are exceptions to the requirement for prior notification to be made 42 days in advance.

Notification of such activities, above the agreed thresholds, will be given at the time the troops involved commence such activities.

Notification will be given in writing of each notifiable military activity in the following agreed form:

A—General Information

- The designation of the military activity;
- The general purpose of the military activity;
- The names of the States involved in the military activity;
- The level of command, organising and commanding the military activity;
- The start and end dates of the military activity.

B—Information on Different Types of Notifiable Military Activities

The engagement of land forces of the participating States in the same exercise activity conducted under a single operational command independently or in combination with any possible air or naval components:

- The total number of troops taking part in the military activity (i.e., ground troops, amphibious troops, airmobile and airborne troops) and the number of troops participating for each State involved, if applicable;
- Number and type of divisions participating for each State;
- The total number of battle tanks for each State and the total number of anti-tank guided missile launchers mounted on armored vehicles;

- The total number of artillery pieces and multiple rocket launchers (100-mm calibre or above);
- The total number of helicopters, by category;
- Envisaged number of sorties by aircraft, excluding helicopters;
- Purpose of air missions;
- Categories of aircraft involved;
- The level of command, organising and commanding the air force participation;
- Naval ship-to-shore gunfire;
- Indication of other naval ship-to-shore support;
- The level of command, organising and commanding the naval force participation.

The engagement of military forces either in an amphibious landing or in a parachute assault by airborne forces in the zone of application for CSBM:

- The total number of amphibious troops involved in notifiable amphibious landings, and/or the total number of airborne troops involved in notifiable parachute assaults;
- In the case of a notifiable amphibious landing, the point or points of embarkation, if in the zone of application for CSBM.

The engagement of formations of land forces on the participating States in a transfer from outside the zone of application for CSBM to arrival points in the zone, or from inside the zone of application for CSBM to points of concentration in the zone, to participate in a notifiable exercise activity or to be concentrated:

- The total number of troops transferred;
- Number and type of divisions participating in the transfer;
- The total number of battle tanks participating in a notifiable arrival or concentration;
- Geographical coordinates for the points of arrival and for the points of concentrations.

C—The Envisaged Area and Timeframe of the Activity

- The area of the military activity delimited by geographic features together with geographic coordinates, as appropriate;
- The start and end dates of each phase (transfers, deployment, concentration of forces, active exercise phase, recovery phase) of activities in the zone of application for CSBM of participating formations, the tactical purpose and corresponding geographical areas (delimited by geographical coordinates) for each phase;
- Brief description of each phase

D—Other Information

- Changes, if any, in relation to information provided in the annual calendar regarding the activity;
- Relationship of the activity to other notifiable activities.

Observation of Certain Military Activities

. . .[Notifiable military] activities will be subject to observation whenever the number of troops engaged meets or exceeds 17,000 troops, except in the case of either an amphibious landing or a parachute assault by airborne forces, which will be subject to observation whenever the number of troops engaged meets or exceeds 5,000 troops. . . .

Each participating State may send up to two observers to the military activity to be observed. . . .

Replies to the invitation will be given in writing not later than 21 days after the issue of invitation. . . .

The host State will not be required to permit observation of restricted location, installations of defence sites.

In order to allow the observers to confirm that the notified activity is non-threatening in character and that it is carried out in conformity with the appropriate provisions of the notification, the host State will:

- at the commencement of the observation programme give a briefing on the purpose, the basic situation, the phases of the activity and possible changes as compared with the notification and provide the observers with a map of the area of the military activity with a scale of 1 to not more than 500,000 and an observation programme with a daily schedule as well as a sketch indicating the basic situation;
- provide the observers with appropriate observation equipment; however, the observers will be allowed to use their personal binoculars, which will be subject to examination and approval by the host State;. . .
- provide opportunities to observe directly forces of the State/States engaged in the military activity so that the observers get an impression of the flow of the activity; to this end, the observers will be given the opportunity to observe major combat units of the participating formations of a divisional or equivalent level and, whenever possible, to visit some units and communicate with commanders and troops; commanders or other senior personnel of participating formations as well as of the visited units will inform the observers of the mission of their respective units;. . .

The participating States need not invite observers to notifiable military activities which are carried out without advance notice to the troops involved unless these notifiable activities have a duration of more than 72 hours. The continuation of these activities beyond this time will be subject to observation while the agreed thresholds for observation are met or exceeded. The observation programme will follow as closely as practically possible all the provisions for observation set out in this document.

Annual Calendars

Each participating State will exchange, with all other participating States, an annual calendar of its military activities subject to prior notification, within the zone of application for CSBM, forecast for the subsequent calandar year. It will be transmitted every year, in writing, through diplomatic channels, not later than 15 November for the following year.

Each participating State will list the above-mentioned activities chronologically and will provide information on each activity in accordance with the following model:

- type of military activity and its designation;
- general characteristics and purpose of the military activity;
- States involved in the military activity;
- areas of the military activity, indicated by appropriate geographic features and/or defined by geographic co-ordinates;

INTERVIEW WITH JACQUES CHIRAC, PRIME MINISTER OF FRANCE, May 16, 1987 (EXCERPTS)[1]

. . .To us, the main thing is that disarmament should go ahead under conditions that will strengthen security. What is the position at present? In the United States, about 10–12,000 nuclear warheads; in the Soviet Union, 10–12,000 that could possibly be dropped on Europe. In Europe, something like 600 nuclear warheads: that's a lot, but nevertheless it is not commensurate. Hence our prime objective, which is a reduction of American and Soviet strategic weapons. We applauded the idea of a 50% cut, which would in fact take us back to the position as it was ten years ago. We should like the process to go further and faster. And then an agreement was reached on removing the medium-range weapons, both the SS-20s and the American euromissiles. Now there is a proposal to go further than that. So we are saying: it is a European issue, not a French one, since, as you know, we are not in NATO and we don't have its missiles on our soil, we don't have any American forces or weapons in France. But on the other hand, being in Europe, we wish our position to be on a footing of solidarity with that of our European neighbours and friends and, naturally, it would worry us if Europe were of a sudden, in a way, totally disarmed between those two gigantic arsenals that can obviously threaten it; and this in a context where, into the bargain, there is in Europe itself a great difference between the Warsaw Pact forces and the NATO and (in France's case) non-NATO forces of the Western democracies. So we are concerned and we say that the question must be carefully handled and thoroughly discussed. Europe's security must be ensured. . . .

HALIFAX STATEMENT ON CONVENTIONAL ARMS CONTROL, May 30, 1987[2]

Within the Alliance, we cherish the ideal that all the peoples of Europe, from the Atlantic to the Urals, should live in peace, freedom and security. To achieve that ideal, bold new steps are required in the field of conventional arms control.

Our objective is the strengthening of stability and security in the whole of Europe, through increased openness and the establishment of a verifiable, comprehensive and stable balance of conventional forces at lower levels.

To work urgently towards the achievement of this objective, we have decided to set up a high level task force on conventional arms control.

It will build on the Western proposals at the CDE conference in Stockholm and at the MBFR negotiations in Vienna, in both of which participating Allied countries are determined to achieve early agreement.

It will take account of Mr Gorbachev's statement of 18 April expressing, in particular, Soviet readiness to pursue conventional force reductions from the Atlantic to the Urals.

An interim report will be presented to the Council in October, and a final report will be discussed at our next meeting in December.

Our aim is a radical improvement in East–West relations in which more confidence, greater openness, and increased security will benefit all.

---

[1] *Survival* (September–October 1987), p. 461.

[2] *Survival* (September–October 1987), p. 466.

CONVENTIONAL ARMS CONTROL: THE WAY AHEAD. STATEMENT ISSUED UNDER THE AUTHORITY OF THE HEADS OF STATE AND GOVERNMENT PARTICIPATING IN THE MEETING OF THE NORTH ATLANTIC COUNCIL IN BRUSSELS, March 2–3, 1988 (EXCERPTS)[1]

At Halifax in 1986, our governments issued a clear call to strengthen stability in the whole of Europe through conventional arms control negotiations. At Brussels later that year they elaborated the basic purposes and methods for such negotiations.

The military confrontation in Europe is the result, not the cause, of the painful division which burdens this continent. While seeking to overcome this division in other ways, we also seek security and stability in Europe at the lowest possible level of armaments. Both arms control and adequate defence programmes can contribute towards this goal.

The Present Realities

(1) The Soviet Union's military presence in Europe, at a level far in excess of its needs for self-defence, directly challenges our security as well as our hopes for change in the political situation in Europe. Thus the conventional imbalance in Europe remains at the core of Europe's security concerns. The problem is to a large extent a function of the Warsaw Pact's superiority in key conventional weapon systems. But it is not only a matter of numerical imbalances. Other asymmetries are also important, for example:

- the Warsaw Pact, based on the Soviet Union's forward–deployed forces, has a capability for surprise attack and large-scale offensive action; the Allies neither have, nor aspire to, such a capability;
- the countries of the Warsaw Pact form a contiguous land mass; those of the Alliance are geographically disconnected;
- the Warsaw Pact can generate a massive reinforcement potential from distances of only a few hundred kilometres; many Allied reinforcements need to cross the Atlantic;
- the Warsaw Pact's military posture and activities are still shrouded in secrecy, whereas those of Allied countries are transparent and under permanent public scrutiny.

(2) These asymmetries are compounded by the dominant presence in Europe of the conventional armed forces of the Soviet Union. They represent 50% of all the active divisions in Europe between the Atlantic and the Urals. This Soviet conventional superiority and its military presence in other Eastern European countries serve a political as well as a military function. They cast a shadow over the whole of Europe.

(3) Conventional arms control is not merely a technical corrective to a self-contained problem. It should be seen in a coherent political and security framework.

A Political and Security Framework

(4) We reiterate our conviction that military forces should only exist to prevent war and to ensure self defence, not for the purpose of initiating aggression and not for the purposes of political or military intimidation. . . .

---

[1]*Survival* (May–June 1988), pp. 275–78.

(5) The relationship between nuclear and conventional forces is complex. The existence of a conventional imbalance in favour of the Warsaw Pact is not the only reason for the presence of nuclear weapons in Europe. The countries of the Alliance are, and will remain, under the threat of Soviet nuclear forces of varying ranges. Although conventional parity would bring important benefits for stability, only the nuclear element can confront a potential aggressor with an unacceptable risk; therefore, for the foreseeable future deterrence will continue to require an adequate mix of nuclear as well as conventional forces.

(6) Hence the determination of our nations to ensure defence preparedness as a means of achieving the stability we seek. We will continue to ensure that our military forces are effective and up-to-date, in particular by:

- continued compliance with the principle of shared risks and responsibilities and acceptance of the priorities essential to the strengthening of our defence capabilities;
- provision of adequate defence expenditure, together with efforts to obtain the greatest return on our defence investment;
- closer co-operation designed to remedy key deficiencies and, in this context, support for recent legislative and other initiatives designed to foster co-operation in the area of conventional armaments, especially research, development, production and procurement;
- helping to meet the needs of the less advantaged Allies in strengthening their conventional defences, thus redressing important existing deficiencies.

(7) . . .In framing their negotiating proposals for conventional stability. . .they will neither make nor accept proposals which would involve an erosion of the Allies' nuclear deterrent capability.

(8) . . .Allied forces are stationed outside their national territory to protect these values and to uphold the solidarity of our free Alliance. They cannot therefore be equated with Soviet forces stationed in Eastern Europe. . . .

(9) . . .both the negotiations which they have now proposed, on conventional stability, as well as those on confidence- and security-building measures, will be undertaken within the framework of the CSCE [Conference on Security and Cooperation in Europe] process.

(10) Those on confidence and security building will involve all 35 CSCE signatory states and will have as their objective to build upon and expand the results of the Stockholm Conference. . . .

(11) At the same time we are conscious of the specific responsibility of the 23 members of the two military alliances in Europe whose forces bear most directly on the essential security relationship in Europe. Hence our decision that distinct and autonomous negotiations on conventional stability should take place between the 23 States.

(12) The adoption of mandates for both of the negotiations must be part of a balanced outcome to the Vienna CSCE Follow-up Meeting, which necessitates substantial progress in all areas of the Helsinki Final Act.

The Allies' Objectives

(13) In accordance with the principles of our approach to conventional arms control, as set out in the Brussels Declaration, our objectives in the forthcoming conventional stability negotiations will be:

- the establishment of a secure and stable balance of conventional forces at lower levels;
- the elimination of disparities prejudicial to stability and security;
- and, as a matter of high priority, the elimination of the capability for launching surprise attack and for initiating large-scale offensive action.

(14) This latter capability is the most worrying in relation to the seizure of territory by an aggressor. Its essential ingredient is the forward deployment of conventional forces capable of rapid mobility and high firepower. Tanks and artillery are among the most decisive components, though other elements of combat capability could prove to be similarly significant. Manpower is also important. But not all items of equipment are appropriate for limitation, if only for technical reasons, and manpower alone is an imprecise guide to offensive capability.

(15) Our aim will be to establish a situation in Europe in which force postures as well as the numbers and deployments of weapon systems no longer make surprise attack and large-scale offensive action a feasible option. We shall pursue this aim on the basis of the following criteria:

- we need to enhance stability in the whole of Europe from the Atlantic to the Urals; and to do so in a way which, while safeguarding the security of all Allies, takes account of the concentrations of Warsaw Pact forces and the particular problems affecting the Central, Southern and Northern regions;
- in seeking to eliminate the ability to conduct large-scale offensive action, we shall focus on the key weapons systems;
- we shall propose provisions dealing with stationed forces, taking account of the weight of forward-developed Soviet conventional forces; we shall also take into consideration capabilities for force generation and reinforcement;
- equal number or percentage reductions by both sides would not eliminate the disparities which threaten stability in Europe. Our proposals will concentrate instead on results and residual entitlements;
- our goal is to redress the conventional imbalance. This can be achieved through a set of measures including, *inter alia*, reductions, limitations, redeployment provisions and related measures as well as the establishment of equal ceilings;
- this outcome will require highly asymmetrical reductions by the East and will entail, for example, the elimination from Europe of tens of thousands of Warsaw Pact weapons relevant to surprise attack, among them tanks and artillery pieces;
- reductions of combat-decisive equipment and modification of the Soviet forward deployment posture will only be part of our approach to reducing the risk of conflict. As a concurrent element in any effort to enhance stability and security, we shall also propose measures to produce greater openness of military activities throughout Europe, safeguard the maintenance of lower force levels, and support a rigorous, effective and reliable monitoring and verification regime;
- this monitoring and verification regime will need to include the exchange of detailed data about forces and deployments; and the right to conduct sufficient on-site inspections to provide confidence that agreed provisions are being complied with.

The Way Ahead

(16) Early agreement on a conventional stability mandate, as part of a balanced outcome to the Vienna Follow-up Meeting of the Conference on Security and Cooperation in

Europe, would be an important step forward. We seek the elimination of the conventional imbalances which so threaten stability and security in Europe. We also seek enhanced respect for human rights and fundamental freedoms on which lasting security and stability ultimately depend.

MIKHAIL GORBACHEV'S ADDRESS TO THE UNITED NATIONS, DECEMBER 7, 1988 (EXCERPTS)[1]

Now let me turn to the main issue—disarmament, without which none of the problems of the coming century can be solved. . . .

Tomorrow marks the first anniversary of the signing of the INF Treaty. I am therefore particularly pleased to note that the implementation of the treaty—the elimination of missiles—is proceeding normally, in an atmosphere of trust and business-like work.

A large breach has thus been made in a seemingly unbreakable wall of suspicion and animosity. We are witnessing the emergence of a new historic reality—a turning away from the principle of overarmament to the principle of reasonable defense sufficiency.

We are present at the birth of a new model of ensuring security—not through the build-up of arms, as was almost always the case in the past, but on the contrary, through their reduction on the basis of compromise.

The Soviet leadership has decided to demonstrate once again its readiness to reinforce this healthy process not only by words but also by deeds.

Today, I can report to you that the Soviet Union has taken a decision to reduce its armed forces.

Within the next two years their numerical strength will be reduced by 500,000 men. The numbers of conventional armaments will also be substantially reduced. This will be done unilaterally, without relation to the talks on the mandate of the Vienna meeting.

By agreement with our Warsaw Treaty allies, we have decided to withdraw by 1991 six tank divisions from the GDR [East Germany], Czechoslovakia and Hungary, and to disband them.

Assault landing troops and several other formations and units, including assault crossing units with their weapons and combat equipment, will also be withdrawn from the groups of Soviet forces stationed in those countries.

Soviet forces stationed in those countries will be reduced by 50,000 men and their armaments, by 5,000 tanks.

All Soviet divisions still remaining in the territory of our allies are being reorganized.

. . .The U.S.S.R. and the United States have built the largest nuclear and missile arsenals. But it is those two countries that, having become specifically aware of their responsibility, were the first to conclude a treaty on the reduction and physical elimination of a portion of their armaments which posed a threat to both of them and to all others.

Both countries possess the greatest and the most sophisticated military secrets. But it is those two countries that have laid a basis for and are further developing a system of mutual verification both of the elimination of armaments and of the reduction and prohibition of their production.

It is those two countries that are accumulating the experience for future bilateral and multilateral agreements.

---

[1]*The Washington Post*, December 8, 1988, p. A32.

We value this. We acknowledge and appreciate the contribution made by President Ronald Reagan and by the members of his administration, particularly Mr. George Shultz.

All this is our joint investment in a venture of historic importance. We must not lose this investment, or leave it idle.

The next U.S. administration headed by President-elect George Bush will find in us a partner who is ready—without long pauses or backtracking—to continue the dialogue in a spirit of realism, openness and goodwill, with a willingness to achieve concrete results working on the agenda which covers the main issues of Soviet–U.S. relations and world politics.

I have in mind, above all, consistent movement toward a treaty on 50 percent reductions in strategic offensive arms while preserving the ABM treaty;

Working out a convention on the elimination of chemical weapons—here, as we see it, prerequisites exist to make 1989 a decisive year;

And negotiations on the reduction of conventional arms and armed forces in Europe.

I also have in mind economic, environmental and humanistic problems in their broadest sense. . . .

Esteemed Mr. President, distinguished delegates,

I am concluding my first address to the United Nations with the same feeling that I had when I began it—a feeling of responsibility to my own people and to the world community.

We are meeting at the end of a year which has meant so much for the United Nations and on the eve of a year from which we all expect so much.

I would like to believe that our hopes will be matched by our joint effort to put an end to an era of wars, confrontation and regional conflicts, to aggressions against nature, to the terror of hunger and poverty as well as to political terrorism.

This is our common goal and we can only reach it together.

Thank you.

STATEMENT ISSUED BY THE NORTH ATLANTIC COUNCIL MEETING IN MINISTERIAL SESSION AT NATO HEADQUARTERS, BRUSSELS, DECEMBER 8–9, 1988[1]

1. In their statement "Conventional Arms Control: The Way Ahead", the Heads of State and Government participating in the meeting of the North Atlantic Council in March 1988 emphasized that the imbalance in conventional forces remains at the core of Europe's security concerns. We shall be presenting specific proposals at the negotiating table to redress this imbalance.

2. We look forward to the early commencement of the two negotiations we have proposed: one on conventional stability between the 23 members of the two military Alliances in Europe and one on confidence- and security-building measures among all 35 signatories of the Helsinki Final Act.

3. In these negotiations we will be guided by:

- the conviction that the existing military confrontation is the result, not the cause, of the painful division of Europe;

[1]NATO Press Service, December 8, 1988.

- the principle of the indivisible security of all our nations. We shall reject calls for partial security arrangements or proposals aimed at separate agreements;
- the hope that the new thinking in the Soviet Union will open the way for mutual agreement on realistic, militarily significant and verifiable arrangements which enhance security at lower levels.

Towards Stability

4. The major threat to stability in Europe comes from those weapons systems which are capable of mounting large-scale offensive operations and of seizing and holding territory. These are above all main battle tanks, artillery and armoured troop carriers. It is in these very systems that the East has such a massive preponderance. Indeed, the Soviet Union itself possesses more tanks and artillery than all the other members of the Warsaw Pact and the Alliance combined. And they are concentrated in a manner which raises grave concerns about the strategy which they are intended to support as well as their role in maintaining the division of Europe.

5. The reductions announced by the Soviet Union are a positive contribution to correcting this situation. They indicate the seriousness with which the conventional imbalances which we have long highlighted as a key problem of European security are now also addressed by the Soviet government. We also welcome the declared readiness of the Soviet Union to adjust their force posture. The important thing is now to build on these hopeful developments at the negotiating table in order to correct the large asymmetries that will still remain and to secure a balance at lower levels of forces. For this, it will be necessary to deal with the location, nationality and the state of readiness of forces, as well as their numbers. Our proposals will address these issues in the following specific ways:

- We shall propose an overall limit on the total holdings of armaments in Europe. This limit should be substantially lower than existing levels, in the case of tanks close to a half. This would mean an overall limit of about 40,000 tanks.
- In our concept of stability, no country should be able to dominate the continent by force of arms. We shall therefore also propose that no country should be entitled to possess more than a fixed proportion, such as 30 percent, of the total holdings in Europe of the 23 participants in each equipment category. In the case of tanks, this would result in an entitlement of no more than about 12,000 tanks for any one country.
- Limiting numbers and nationality of forces would not by itself affect the stationing of forces on other countries' territory. Stationed forces, particularly those in active combat units, are especially relevant to surprise attack. We shall propose limits on such forces.
- Our proposal will apply to the whole of Europe. In order to avoid undue concentration of these weapon categories in certain areas of Europe, we shall propose appropriate sub-limits.

6. To buttress the resulting reductions in force levels in the whole of Europe, we shall propose stabilizing measures. These could include measures of transparency, notifications and constraint applied to the deployment, movement and levels of readiness of conventional armed forces, which include conventional armaments and equipment.

7. Finally, we shall require a rigorous and reliable regime for monitoring and verification. This would include the periodic exchange of detailed data about forces and deployments, and the right to conduct on-site inspections.

Towards Transparency

8. Greater transparency is an essential requirement for real stability. Therefore, within the framework of the CSCE process, the negotiations on confidence- and security-building measures form an essential complement to those on conventional stability. We are encouraged thus far by the successful implementation of the Stockholm Document and we consider that the momentum must be maintained.

9. In order to create transparency of military organization, we plan to introduce a proposal for a wide-ranging, comprehensive annual exchange of information concerning military organization, manpower and equipment as well as major weapon deployment programmes. To evaluate this information we will propose modalities for the establishment of a random evaluation system.

10. In addition, in order to build on the success of the Stockholm Document and to create greater transparency of military activities, we will propose measures in areas such as:

- more detailed information with regard to the notification of military exercises,
- improvements in the arrangements for observing military activities,
- greater openness and predictability about military activities,
- a strengthening of the regime for ensuring compliance and verification.

11. Finally, we shall propose additional measures designed to improve contacts and communications between participating states in the military field; to enhance access for military staffs and media representatives; and to increase mutual understanding of military capabilities, behaviour and force postures. We will also propose modalities for an organized exchange of views on military doctrine tied to actual force structures, capabilities and dispositions in Europe.

A Vision for Europe

12. We will pursue these distinct negotiations within the framework of the CSCE process, because we believe that a secure peace cannot be achieved without steady progress on all aspects of the confrontation which has divided Europe for more than four decades. Moreover, redressing the disparity in conventional forces in Europe would remove an obstacle to the achievement of the better political relationship between all states of Europe to which we aspire. Conventional arms control must therefore be seen as part of a dynamic process which addresses the military, political, and human aspects of this division.

13. The implementation of our present proposals and of those we are making for further CSBMs [confidence- and security-building measures] will involve a quantum improvement in European security. We will wish to agree and implement them as soon as possible. In the light of their implementation we would then be willing to contemplate further steps to enhance stability and security in Europe, for example:

- further reductions or limitations of conventional armaments and equipment;
- the restructuring of armed forces to enhance defensive capabilities and further reduce offensive capabilities.

Our vision remains that of a continent where military forces only exist to prevent war and to ensure self defence, not for the purpose of initiating aggression or for political or military intimidation.

MANDATE FOR NEGOTIATION ON CONVENTIONAL ARMED FORCES IN EUROPE, JANUARY 10, 1989[1]

Signed on 10 January 1989 at Palais Liechtenstein, Vienna by the representatives of Belgium, Bulgaria, Canada, Czechoslovakia, Denmark, France, the Federal Republic of Germany, the German Democratic Republic, Greece, Hungary, Iceland, Luxembourg, the Netherlands, Norway, Poland, Portugal, Romania, Spain, Turkey, the Union of Soviet Socialist Republics, the United Kingdom and the United States of America held consultations in Vienna from 17 February 1987 to 10 January 1989.

These States . . . Determined that a Negotiation on Conventional Armed Forces in Europe should take place in the framework of the CSCE process . . .

Have agreed on the following provisions: . . .

Objectives and Methods

The objectives of the negotiation shall be to strengthen stability and security in Europe through the establishment of a stable and secure balance of conventional armed forces, which include conventional armaments and equipment, at lower levels; the elimination of disparities prejudicial to stability and security; and the elimination, as a matter of priority, of the capability for launching surprise attack and for initiating large-scale offensive action. Each and every participant undertakes to contribute to the attainment of these objectives.

These objectives shall be achieved by the application of militarily significant measures such as reductions, limitations, redeployment provisions, equal ceilings, and related measures, among others. In order to achieve the above objectives, measures should be pursued for the whole area of application with provisions, if and where appropriate, for regional differentiation to redress disparities within the area of application and in a way which precludes circumvention.

The process of strengthening stability and security should proceed step-by-step, in a manner which will ensure that the security of each participant is not affected adversely at any stage.

Scope and Area of Application

The subject of the negotiation shall be the conventional armed forces, which include conventional armaments and equipment, of the participants based on land within the territory of the participants in Europe from the Atlantic to the Urals.

The existence of multiple capabilities will not be a criterion for modifying the scope of the negotiation:

> No conventional armaments for equipment will be excluded from the subject of the negotiation because they may have other capabilities in addition to conventional ones. Such armaments or equipment will not be singled out in a separate category;
> Nuclear weapons will not be a subject of this negotiation.

Particular emphasis will initially be placed on those forces directly related to the achievement of the objectives of the negotiation set out above.

Naval forces and chemical weapons will not be addressed.

---

[1]*Survival* (May–June 1989), pp. 274–75.

The area of application[1] shall be the entire land territory of the participants in Europe from the Atlantic to the Urals, which includes all the European island territories of the participants. In the case of the Soviet Union the area of application includes all the territory lying west of the Ural River and the Caspian Sea. In the case of Turkey the area of application includes the territory of Turkey north and west of the following line: the point of intersection of the border with the 39th parallel, the Muradiye, Patnos, Karayazi, Tekman, Kemaliye, Feke, Ceyhan, Dogankent, Gozne and then to the sea.

Exchange of Information and Verification

Compliance with the provisions of any agreement shall be verified through an effective and strict verification regime which, among other things, will include on-site inspections as a matter of right and exchanges of information.

Information shall be exchanged in sufficient detail so as to allow a meaningful comparison of the capabilities of the forces involved. Information shall also be exchanged in sufficient detail so as to provide a basis for the verification of compliance.

The specific modalities for verification and the exchange of information, including the degree of detail of the information and the order of its exchange, shall be agreed at the negotiation proper.

Procedures and Other Arrangements

. . . The participants decided to take part in meetings of the States signatories of the Helsinki Final Act to be held at least twice during each round of the Negotiation on Conventional Armed Forces in Europe in order to exchange views and substantive information concerning the course of the Negotiation on Conventional Armed Forces in Europe. . . .

The participants will take into consideration the views expressed in such meetings by other CSCE participating States concerning their own security. Participants will also provide information bilaterally.

The participants undertake to inform the next CSCE Follow-up Meeting of their work and possible results and to exchange views, at that meeting, with the other CSCE participating States on progress achieved in the negotiation. . . .

POSITION PAPER PROVIDED BY THE DELEGATIONS OF BELGIUM, CANADA, DENMARK, FEDERAL REPUBLIC OF GERMANY, FRANCE, GREECE, ICELAND, ITALY, LUXEMBOURG, NETHERLANDS, NORWAY, PORTUGAL, SPAIN, TURKEY, UNITED KINGDOM AND UNITED STATES, MARCH 6, 1989[2]

Objectives

(1) The objectives of these negotiations as agreed in the mandate are:

> the establishment of a secure and stable balance of conventional forces at lower levels;
> the elimination of disparities prejudicial to stability and security;

[1]The participants will be guided by the language on non-circumvention as set out in the section on Objectives and Methods.

[2]*Survival* (May–June 1989), pp. 275–76.

> the elimination, as a matter of high priority, of the capability for launching surprise attack and for initiating large-scale offensive action.

(2) Through the approach outlined below the Western Delegations will seek to establish a situation in which surprise attack and large-scale offensive action are no longer credible options. . . .

Rationale

(3) The rationale for our approach is as follows:

> the present concentration of forces in the area from the Atlantic to the Urals is the highest ever known in peace-time and represents the greatest destructive potential ever assembled. Overall levels of forces, particularly those relevant to surprise attack and offensive action such as tanks, artillery and armoured troop carriers, must therefore be radically reduced. It is the substantial disparity in the numbers of these systems, all capable of rapid mobility and high firepower, which most threatens stability in Europe. These systems are also central to the seizing and holding of territory, the prime aim of any aggressor;
> no one country should be permitted to dominate Europe by force of arms: no participant should therefore possess more than a fixed proportion of the total holdings of all participants in each category of armaments, commensurate with its needs for self defense;
> addressing the overall number and nationality of forces will not by itself affect the stationing of armaments outside national borders: additional limits will also be needed on forces stationed on other countries' territory;
> we need to focus on both the levels of armaments and state of readiness of forces in those areas where the concentration of such forces is greatest, as well as to prevent redeployment of forces withdrawn from one part of the area of application to another. It will therefore be necessary to apply a series of interlocking sub-limits covering forces throughout the area, together with further limits on armaments in active units.

(4) The following specific measures within the area of application would fulfill these objectives:

Rule 1: Overall Limit

The overall total of weapons in each of the three categories identified above will at no time exceed:

- main battle tanks 40,000
- artillery pieces 33,000
- armoured troop carriers 56,000

Rule 2: Sufficiency

No one country may retain more than 30 per cent of the overall limits in these three categories, i.e.

- main battle tanks 12,000
- artillery pieces 10,000
- armoured troop carriers 16,800

Rule 3: Stationed Forces

Among countries belonging to a treaty of Alliance neither side will station armaments outside national territory in active units exceeding the following levels:

- main battle tanks 3,200
- artillery pieces 1,700
- armoured troop carriers 6,000

Rule 4: Sub-limits

In the areas indicated below, each group of countries belonging to the same treaty of Alliance shall not exceed the following levels:
(1) In the area consisting of Belgium, Denmark, the Federal Republic of Germany, France, Greece, Iceland, Italy, Luxembourg, the Netherlands, Norway, Portugal, Spain, Turkey, the United Kingdom, Bulgaria, Czechoslovakia, the German Democratic Republic, Hungary, Romania and the territory of the Soviet Union west of the Urals comprising the Baltic, Byelorussian, Carpathian, Moscow, Volga, Urals, Leningrad, Odessa, Kiev, Trans-Caucasus, North Caucasus military districts:

- main battle tanks 20,000
- artillery pieces 16,500
- armoured troop carriers 28,000 (of which no more than 12,000 AIFV)

(2) In the area consisting of Belgium, Denmark, the Federal Republic of Germany, France, Italy, Luxembourg, the Netherlands, Portugal, Spain, the United Kingdom, Czechoslovakia, the German Democratic Republic, Hungary, Poland and the territory of the Soviet Union west of the Urals comprising the Baltic, Byelorussian, Carpathian, Moscow, Volga, Urals military districts in active units:

- main battle tanks 11,300
- artillery 9,000
- armoured troop carriers 20,000

(3) In the area consisting of Belgium, Denmark, the Federal Republic of Germany, France, Italy, Luxembourg, the Netherlands, the United Kingdom, Czechoslovakia, the German Democratic Republic, Hungary, Poland and the territory of the Soviet Union comprising the Baltic, Byelorussian, Carpathian military districts in active units:

- main battle tanks 10,300
- artillery 7,600
- armoured troop carriers 18,000

(4) In the area consisting of Belgium, the Federal Republic of Germany, Luxembourg, the Netherlands, Czechoslovakia, the German Democratic Republic and Poland in active units:

- main battle tanks 8,000
- artillery 4,500
- armoured troop carriers 11,000

(5) Rule 4 is to be seen as an integrated whole which will only be applied simultaneously and

across the entire area from the Atlantic to the Urals. It will be for the members of each Alliance to decide how they exercise their entitlement under all of these measures.

Rule 5: Information Exchange

Each year holdings of main battle tanks, armoured troop carriers and artillery pieces will be notified, disaggregated down to battalion level. This measure will also apply to personnel in both combat and combat support units. Any change of notified unit structures above battalion level, or any measure resulting in an increase of personnel strength in such units, will be subject to notification, on a basis to be determined in the course of the negotiations.

Measures for Stability, Verification and Non-Circumvention

(5) As an integral part of the agreement, there would be a need for:

> stabilizing measures: to buttress the resulting reductions in force levels in the ATTU area. These should include measures of transparency, notification and constraint applied to the deployment, movement, and levels of readiness of conventional armed forces which include conventional armaments and equipment;
> verification arrangements: to include the exchange of detailed data about forces and deployments, with the right to conduct on-site inspection, as well as other measures designed to provide assurance of compliance with the agreed provisions;
> non-circumvention provisions: inter alia, to ensure that the manpower and equipment withdrawn from any one area do not have adverse security implications for any participating state;
> provision for temporarily exceeding the limits set down in Rule 4 for pre-notified exercises.

The Longer Term

(6) In the longer term, and in the light of the implementation of the above measures, we would be willing to contemplate further steps to enhance stability and security in Europe, such as:
further reductions or limitations of conventional armaments and equipment;
the restructuring of armed forces to enhance defensive capabilities and further to reduce offensive capabilities.

CONCEPTUAL FRAMEWORK OF AGREEMENT ON CONVENTIONAL ARMED FORCES IN EUROPE PROPOSED BY BULGARIA, CZECHOSLOVAKIA, THE GERMAN DEMOCRATIC REPUBLIC, HUNGARY, POLAND, ROMANIA, AND THE UNION OF SOVIET SOCIALIST REPUBLICS, MARCH 9, 1989[1]

Objectives

(1) The objectives of the Agreement will be: to strengthen stability and security in Europe through deep cuts in conventional armed forces, including conventional armaments and

[1]*Survival* (May–June 1989), pp. 278–81.

equipment, of the Warsaw Treaty member States and of the NATO member countries so as to establish in this way a balance at lower levels at which both military alliances will keep forces and systems necessary solely for defense and insufficient to launch surprise attack or conduct offensive operations; to restructure and redeploy their armed forces on strictly defensive principles.

(2) The process of strengthening stability and security on the European continent should proceed stage by stage and in a manner that will not upset the overall balance or prejudice anyone's security at all stages of the negotiations.

(3) These objectives will be achieved through reductions, limitations, appropriate redeployment measures, equal collective ceilings on armed forces and conventional armaments both throughout the European zone and in its individual regions.

The scope and procedure of reduction of national and foreign troops down to agreed levels would be decided upon within each alliance on the basis of the principles and criteria to be agreed at the negotiation.

(4) The reductions in conventional armed forces and armaments will be effected on the basis of reciprocity, with all the Participating States without exception making their appropriate contributions with account taken of their military potentials.

The reductions will be accompanied by corresponding cuts in military expenditures and by measures to convert their military production.

(5) All the participants will refrain from steps running counter to the above objectives and will not build up their conventional armed forces and armaments.

(6) These objectives can be achieved both through agreed steps and unilateral measures to reduce their armed forces and armaments.

Zone of Application of Agreement

The zone of application of agreement will be the entire land territory of the Participating States in Europe from the Atlantic to the Urals, including a part of the Asian territory of Turkey and the Soviet Transcaucasus as well as all the European island territories of the Participating States including the Faroe Islands, Svalbard, the Islands of Azores and Madeira, the Canary Islands, Franz Josef Land and Novaya Zemlya.

The First Stage (no later than 1991–1994)

(1) At the first stage of reductions all the Participating States will eliminate imbalances and asymmetries between NATO and the Warsaw Treaty as regards both troop numbers and main armaments and will make steps to eliminate the capability to launch surprise attack or initiate large-scale offensive action.

(2) For this purpose attention will be focused on reducing the most destabilizing types and categories of armaments such as attack combat aircraft of short-range/tactical aviation, tanks, combat helicopters, combat armoured vehicles and armoured personnel carriers, artillery including multiple launch rocket systems and mortars.

(3) The Participating States will reduce their conventional armed forces and armaments down to equal collective ceilings that would be 10–15% lower than the lowest levels possessed by the military political alliances. These ceilings will be agreed between the Participating States in absolute terms. . . .

The armaments and equipment being reduced are eliminated under agreed procedures at specially assigned locations or are converted for civilian use. A part of the armaments and equipment are put in temporary storage under international control.

Surprise Attack Prevention Measures

(1) Starting with the first stage onwards strips (zones) of reduced levels of armaments would be established along the line of contact between the two military political alliances, from where the most dangerous destabilizing types of conventional armaments and equipment would be pulled out, reduced, or limited and where limitations would be introduced on military activities.

The establishment of such zones in Central Europe and in other regions of the European continent could be effected on the basis of the existing and possible new proposals.

(2) The depth of the strips (zones) of reduced levels of armaments could be agreed with account taken of geographical factors and performance characteristics of the main types of armaments.

Confidence- and security-building measures would limit military activities within the strips (zones) and would, appropriately, provide for an increasingly rigorous regime as the line of contact is approached. In particular, they would affect the scope and number of concurrent exercises, the duration and frequence of exercises, and envisage a ban on large-scale exercises and limitations on troop transfers.

The Second Stage (1994–1997)

. . . (2) At the second stage the armed forces of each side would be cut approximately by 25% (by about 500,000 men) with their organic armaments.

(3) Along with further substantial lowering of the levels of the most destabilizing types and categories of armaments the Participating States shall take steps to reduce other categories of armaments not affected by the first-stage cuts.

(4) The Participating States will make further steps to restructure their armed forces on the basis of the principle of sufficiency for defense.

(5) They will elaborate and make agreed steps to develop predictability and openness in day-to-day military activities and to lower their levels.

The Third Stage (1997–2000)

(1) During the third stage further reductions of armed forces and conventional armaments shall be implemented and the armed forces of the participating States will be given a strictly defensive character.

(2) The Participating States will reach agreement on ceilings on all other categories of armaments.

(3) The Participating States will reach agreements on the principles of armed forces development by which they will abide in the future for the purposes of maintaining a secure and stable peace in Europe.

Verification

(1) The Participating States will agree to exchange data regarding manpower strength, number of conventional armaments, and deployment of military formations and to verify them, including through on-site inspections.

(2) There would be envisaged the establishment of a comprehensive and effective system of verification of compliance with agreements including land and air on-site inspections without the right to refuse. There would be created checkpoints to monitor entry/exit both along and

inside the strips (zones) of reduced levels of armaments and in the reduction area (at railway stations, junctions, airfields, ports). Such technical means of verification as artificial Earth satellites, aircraft, helicopters, ground automatic recording systems, including the ones developed through international cooperation, could also be used for the purposes of verification.

(3) There would be verification of the process of reduction, elimination (dismantlement, conversion) and storage of armaments, disbandment of formations and units, non-excess of the strength of armed forces and the number of armaments as well as of the activities of the troops remaining after reductions.

(4) An international verification (consultative) commission would be set up and given wide powers (observation, inspection, consideration of disputes, etc.), which would be made up of representatives of the participating States.

(5) A prominent part in the implementation of verification and control measures should be played by the highest representative bodies—Parliaments, National Assemblies and the Supreme Soviet which could act as guarantors of the reductions and redeployment of the armed forces and conventional armaments of the appropriate countries. Related matters could be discussed within the framework of foreign and military affairs committees and reflected in appropriate statements to be made on behalf of the parliaments. . . .

SPEECH BY HANS-DIETRICH GENSCHER, FRG MINISTER FOR FOREIGN AFFAIRS, IN THE GERMAN BUNDESTAG, BONN, APRIL 27, 1989[1]

Our debate is taking place at a time of movement in Europe. It is evident that the people of Europe are separated, but Europe is indivisible. The categorical imperative of the European democracies is: first, to continue resolutely the process of unification within the European Community; second, to overcome the separation in Europe through cooperation, through respect for human rights, through disarmament agreements and through confidence-building; third, never to forget that both the foregoing are only possible on the basis of a secure defense capacity within a Western alliance capable of action.

A peaceful order in Europe in which borders cease to be of a divisive nature, in which nations can live together without mutual fear, in which they can determine their own political and social systems—such a peaceful order is no longer merely a vision. The prospects for realizing it have improved. We must develop further and strengthen the cooperative elements of coexistence. The turn of the tide in European international politics is irreversible and unmistakable. Nothing is more powerful than an idea whose time has come. This is the idea of eliminating hostility from international relations. It is the idea of demilitarizing East–West relations. It is the idea of deideologizing East–West relations. It is the idea of dialogue and cooperation, the idea of developing new peace structures. These are the topics for the forthcoming summit of the Western alliance.

The central question for the West is whether it regards the democratization and reform in the socialist countries as a danger or as an opportunity that it is willing to make use of. The answer can only be: This is a historic opportunity. We must not let it slip by, nor idly look on from afar, but must seek to exercise creative influence. This is our responsibility! The European Community is the core of Europe's future structure. It is a living model for a peaceful European order, with Franco–German cooperation as its centerpiece. Only with a

---

[1]*Statements and Speeches*, vol. XII, no. 10, German Information Center, New York, April 28, 1989.

dynamic European Community can a new and lasting peaceful order emerge in the whole of Europe. The increasing appeal, the attraction and fascination of our liberal model, which has proved successful in social terms, are now evident. There is certainly no cause for us to be faint-hearted or anxious, but to be confident and enterprising. But only with a Western alliance capable of action will we be able to take advantage without risk, of the new opportunities arising in West–East relations.

Who would deny that the elections in the Soviet Union have shown how deeply the goal of democratization is rooted in the hearts and minds of people in the Soviet Union? The election results were not just a rejection of the opponents to reform in the Soviet Union. They were also a rejection of those in the West who claim that perestroika is merely a game for intellectuals. No, people worldwide want freedom. They make use of the slightest opportunity for gaining it. Nobody has to learn freedom. This is the recognition gained. This process of reform in the Soviet Union is a manifestation of European consciousness. Realistic prospects for fundamentally reshaping international relations and setting up a peaceful order in Europe, as demanded by the Alliance's Harmel Report, are becoming visible.

The INF Treaty showed that disarmament is possible, that it is verifiable and that it leads to greater security. The Vienna negotiations on conventional stability can be assisted by new efforts at the Geneva negotiations for the global prohibition of chemical weapons. The time is ripe for this.

Disarmament must not exclude any class of weapons. The comprehensive concept to be adopted at Brussels in May is called a "concept of arms control and disarmament," and not a "concept of armament." Why should it therefore exclude the demand for negotiations already voiced in 1987 and 1988? I feel we have a great deal to fear in Europe and worldwide, but we certainly need not fear negotiations or a Soviet Union that is willing to negotiate. Whoever seeks disarmament must negotiate on disarmament. Disarmament cannot be obtained by opposing the other side, but only by negotiating with it. And a realistic policy of disarmament requires that disarmament steps be made irreversible through negotiations and treaties so that there is no reversion to a new arms race.

The Federal Republic of Germany renders a significant contribution to common Western security—it makes the main contribution to conventional defense. We owe it to the people of our country and to the members of the Bundeswehr, who serve the cause of peace and freedom, to make actual use of absolutely every opportunity for achieving greater security through disarmament. The great goal of the Western alliance is to prevent war. This is the purpose of our common strategy. We endorse the Alliance's observation that at present there is no foreseeable alternative to the concept of preventing war by deterrence based on a suitable mix of adequate and effective nuclear and conventional forces. In the case of nuclear forces, land, sea and air-based systems are, under the existing circumstances, needed in Europe, too. But we also know that the shorter the ranges, the less nuclear weapons are suited for preventing war.

In accordance with our Alliance's inalienable doctrine, nuclear weapons serve the political goal of preventing war. Any other interpretation would lead us into a war-fighting scenario, which would mean the end of all Europe. It is therefore essential to create an additional network of cooperative security which prevents even more reliably a conventional or a nuclear war. Nor do we want a so-called limited nuclear war. In this respect, too, there must be no zones of differing security—neither in the Alliance nor in Europe. The East's conventional superiority continues to lie at the heart of the security problems in Europe.

To replace this by conventional stability with fewer weapons, by eliminating the capability for surprise attack and for large-scale offensive action, that is the objective of the Vienna negotiations. And that is why, with regard to the question of whether or not the

Federal Government deems it necessary to have a follow-up system for the Lance short-range missile in the second half of the 90s, it is crucial for us to achieve agreements of that kind as well as the other objectives mentioned in the policy statement. No one can today reliably predict what the political and security situation will be like in 1992. And that is why it is also not possible today to decide whether a decision must or not be taken in 1992 for the deployment of such a follow-up system in 1996.

With its conscripted army, the Bundeswehr, and by virtue of the fact that our security policy is accepted by the people, the Federal Republic is rendering an indispensable contribution to the security of all the allies. And this entitles us to have a big say in the decision-making process, including Alliance decisions on matters of defense, arms control and disarmament. Our Alliance of free democracies is capable of achieving a consensus on all matters on the basis of mutual respect. That is what we want. It is no sign of weakness when democratic nations try to agree on a common policy through dialogue. As the Chancellor has pointed out, we have no need to provide further evidence of our determination to do everything necessary to maintain common security. We have done so with the significant decisions we made in 1979 and 1983.

But at the same time we expect all our partners to stand by our side as we seek to remove the East's superiority in short-range missiles through negotiation. And we also expect them to show understanding when we labor over the decision as to whether or not it is necessary to deploy new short-range nuclear missiles. In no other category of weapons is Eastern Europe's superiority so great. It is therefore in the West's fundamental security interests to remove that superiority by negotiating equal ceilings as already called for in Reykjavik.

The Federal Republic of Germany shares all, I repeat all, present security risks with its allies. But the threat emanating from Soviet short-range missiles affects us in particular. They do not have the range to reach the territory of most of our partners. So not only is it understandable that we should want to eliminate the Soviet superiority in this category of weapons by negotiation, it is in keeping with the Alliance's basic philosophy of creating more stability by removing superiorities.

In making our decision on new short-range nuclear weapons we are talking about systems which can reach the territory of Poland and Czechoslovakia, nations who suffered so terribly during the Second World War. And we are talking about short-range nuclear weapon systems which can reach the other part of our fatherland. Thus, if we are called upon to make such a decision we will not forget this fact, and I state this here on my own personal responsibility. The members of the Federal Government have sworn on oath to dedicate their efforts to the well-being of the German people. The obligation deriving from that oath does not stop at the border cutting through Germany. The responsibility for the nation established by the oath does not exclude my native region, the town where I was born, nor the people in the GDR. Indeed, it includes them. Our sincerity with regard to the German nation is not to be judged by what is said in fine-sounding speeches but by our efforts, day in, day out, to strengthen peace and stability and enhance human rights in Europe, and by our efforts to promote cooperation and disarmament. The requirement of our Basic Law to promote peace applies to our relations with all European nations. We have proved our determination to do everything necessary to preserve freedom and security. But we shall seize every opportunity for cooperation, detente and disarmament with the same determination.

I appeal to our American friends, to whom we owe so much, whose airlift to save Berlin we remember vividly, and whose contribution to Europe's security is indispensable, not to worry about our pausing to think. They would only have cause to worry if we again became unscrupulous. Pausing to think before making decisions on new nuclear weapons is an expression of responsibility. It is anything but a sign of weakness. And it should not be criticized as such. It is a credit to the citizens of our country and it is a credit to all those who are called upon to exercise political responsibility.

We Germans, this we say to all our friends, are not trying to follow our own separate road into the future. By opting for democracy, the Western Alliance and the European Community, we have irrevocably committed ourselves. We are very much aware that any attempt to detach ourselves from that community, to separate Germany's fate from that of the whole of Europe, to "de-Europeanize" it, would be inimical to the mission prescribed by our constitution: to foster peace in Europe. It would in fact lead to our isolation in both East and West. Through our membership of the community of Western democracies we have used responsibly the freedom we regained on 8 May 1945. By taking that step we entered into the most far-reaching union possible for a state: the union based on fundamental values.

On the other hand, we Germans also have neighbors who do not belong to this community of Western democracies but who suffered during the Second World War, neighbors who likewise desire peace. We Germans do not wish to, nor can we forget, what happened to the Polish people, nor the suffering and sacrifices of the peoples of the Soviet Union. The sorrowful experience of this century makes these nations watch us attentively. And if it is true that East and West need a bridge of trust, then we Germans must build its main support. It is neither presumption nor arrogance on our part but rather a deep awareness of our historical task when we Germans feel we have a special responsibility for confidence-building between East and West, and act accordingly.

This century the Germans have pursued quite different aims from that of good neighborliness with all Europeans. Today, after two murderous world wars and after decades of confrontation between East and West, Europe has a historic opportunity to embark wholeheartedly on the task of establishing a lasting peaceful order in Europe. We Germans have a special role to play in accomplishing this task. The French diplomat and poet Paul Claudel wrote in 1945: "Germany is not there to divide nations but to let all the different nations around it feel that they cannot live without one another." We shall live up to this responsibility for peace.

PRESIDENT GEORGE BUSH'S STATEMENT ON THE NEW ARMS PLAN, BRUSSELS, MAY 29, 1989[1]

I'll have a brief statement before taking some questions. This morning I met with the other NATO leaders and shared with them my views on the role of the North Atlantic alliance in a changing Europe. NATO, we all agree, is one of the great success stories, and it's guaranteed the peace in Europe, provided a shield for 40 years of freedom and prosperity.

And, now, our alliance faces new challenges at a time of historic transition as we seek to overcome the division of Europe. I call it "beyond containment."

And today I am proposing a major initiative to help move us toward that momentous objective. If it were accepted, it would be a revolutionary conventional arms control agreement.

I believe the alliance should act decisively now to take advantage of this extraordinary opportunity and I urge that NATO adopt a four-point proposal to bring the Vienna negotiations to a speedy conclusion.

First, lock in Eastern acceptance of the proposed Western ceilings on each side's

---

[1]*The New York Times,* May 30, 1989, p. A13.

holdings of tanks and armored troop carriers. Additionally, we would seek agreement on a similar ceiling for artillery, provided there are some definitional questions that have to be resolved there. But all of the equipment reduced would be destroyed. . . . We would then . . . expand our current NATO proposal so that each side would reduce to 15 percent below current NATO levels in two additional categories: Attack and assault, or transport, helicopters; and all land-based combat aircraft. All of the equipment reduced would be destroyed.

And third, propose a 20 percent cut in combat manpower in United States-stationed forces and a resulting ceiling in U.S. and Soviet ground and air forces stationed outside of national territory in the Atlantic to the Ural zone at approximately 275,000 each. This manpower ceiling will require the Soviets to reduce their forces in Eastern Europe by about 325,000 people. Withdrawn soldiers and airmen on both sides would be demobilized.

And then, fourthly, accelerate the timetable for reaching a C.F.E. agreement along these lines and implementing the required reductions. I believe it should be possible to reach such agreement in six months or maybe a year and to accomplish the reductions by 1992 or 1993. . . .

Now, if the Soviet Union accepts this fair offer, the results would dramatically increase stability on the continent and transform the military map of Europe.

We can and must begin now to set out a new vision for Europe at the end of this century.

This is a noble mission that I believe the alliance should be ready to undertake. And I have no doubt that we are up to the task. And, incidentally, in addition to these arms control proposals I mentioned in there that we are prepared to change our no-exceptions policy on trade. And I called again for a ban on chemical weapons. And I would reiterate my support for our open-skies proposal and in the meeting it was discussed by the President—the Prime Minister of Canada.

About the Authors

WALTER LAQUEUR is chairman of the International Research Council of the Center for Strategic and International Studies in Washington, D.C., as well as director of the Institute of Contemporary History in London. He is the editor of *The Washington Quarterly* and has been the editor of the *Journal of Contemporary History* since 1965. He has written widely about West and East European affairs. Among his books are *Europe Since Hitler*, a history of Europe since World War II, which has been published in many editions; *Russia and Germany*; and most recently *The Long Road to Freedom: Russia and Glasnost*.

LEON SLOSS is president of Leon Sloss Associates, Inc., a consulting firm specializing in political-military affairs. He spent twenty years in the U.S. government in a variety of positions dealing with political-military affairs and defense strategy. He has served as Acting Director, Arms Control and Disarmament Agency. He has lectured on defense policy at the Army War College, the Foreign Service Institute, and the National Defense University, and under the auspices of the United States Information Agency, he has also lectured in Israel, Japan, and throughout Europe. He is coeditor of *A Game for High Stakes: Lessons Learned in Negotiating with the Soviet Union*.

PHILIPP BORINSKI recently received his Ph.D. in Comparative Government from Georgetown University. His dissertation was "Neutralism in the West German Peace Movement: The Question of the Evaluative Background." He is a reserve lieutenant in the West German Army.

Index